YOU –

COAX –

AND SMITH

By K3MT

Culpeper, VA

FIRST EDITION - JULY 2017

Copyright © 2017 by Michael J. Toia

All rights reserved. Reproduction of any or all of this work in any manner, without the expressed written permission of the author, is prohibited.

No liability is assumed with respect to use or application of the information herein.

ISBN: 978-0-9600859-3-4

JOKALYM PUBLISHING

16206 Glenhollow Ct   Culpeper, VA 22701   k3mt@arrl.net

PREFACE

The smith chart, developed by Phillip H. Smith, was published in the January 1939 issue of *Electronics* magazine. Smith was employed by Bell Labs, according to references available today. The chart as an aid in understanding and designing antennas first came to my attention at age 16 by way of my various mentors. A radio amateur, I found the discipline of antenna engineering fascinating, and that has continued to this day as I enter octogenarianship.

As a recent retiree, my last several years were dedicated to building a School of Antennas and Propagation as a conduit to share sixty years of antenna engineering experience with younger staff members. That assignment has been particularly rewarding, with this book being but one offshoot of the effort. In it is shown the smith chart, how to use it for designing transmission line systems, and how to draw it from memory on a blank piece of paper at any time as may be required.

Author's Notes

Tho the term *coax* appears in this work's title, the information herein is about RF transmission lines used in radio work. Coax is but one type of line. Another common type is the two-parallel wire feedline, the most familiar material being 300 ohm "twin lead" once generally used for television antennas. Results herein apply to all such lines.

Coax *per se* is available in 50 ohm, 75 ohm, and rarely 93 ohm material plus others. Twin lead is available in 75 ohm, 300 ohm, and 450 ohm cables, and more. The common lamp cord material sold in hardware stores can be used as 75 ohm cable for dipole antennas. And each of the four twisted pairs in Ethernet cable has an impedance of 100 ohms. Home- or field-made transmission lines, often a pair of wires separated by insulators every foot or so, also appear from time to time, as does a single wire above ground – with an impedance near 500 ohms, more or less.

To Doctor Moore the Elder, mathematics department, Carnegie Tech, my undergraduate math professor 1955 – 1959. His enthusiasm drew our class deeply into the wonderful world of mathematics, the tools of physics. To this day I yet recall his various math quips such as: "Take an epsilon … take two – they're small."

INDEX

1. AC / DC

2. COAX AND ITS IMPEDANCE

3. DRAWING A SMITH CHART

4. FAMILIARIZATION

5. IMPEDANCE TRANSFORMER
    RESISTIVE LOADS

6. IMPEDANCE TRANSFORMER
    GENERAL LOADS
    ANTENNA TUNING

7. CABLE BALUNS

8. ZEPPS AND J-POLES

9. MONOPOLE AND DIPOLE SHUNTS

10. CONCEPTS OF BROADBANDING

11. Z_o

A. APPENDIX A – PHASOR ARITHMETIC

B. APPENDIX B – THE $z - \rho$ CONFORMAL MAPPING

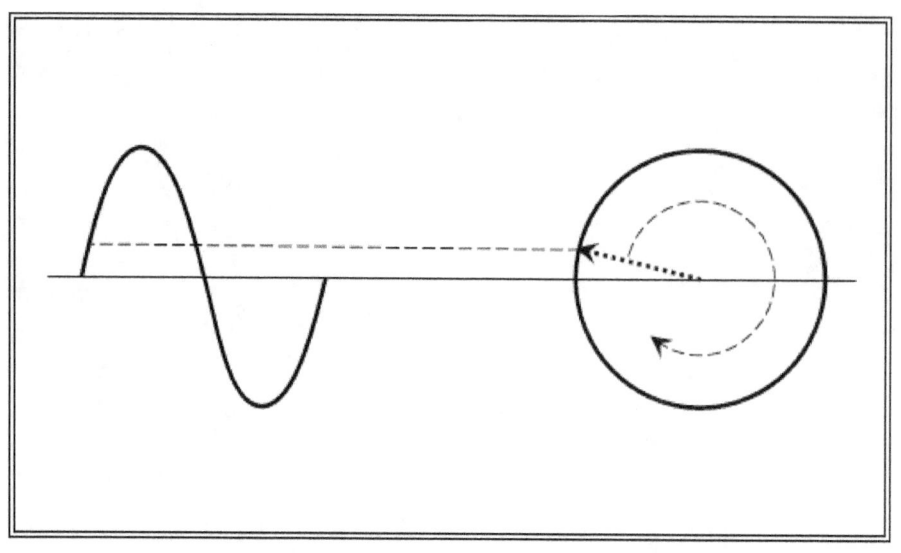

SPINNING PHASOR REPRESENTATION OF A SINE WAVE
FIGURE 1-1

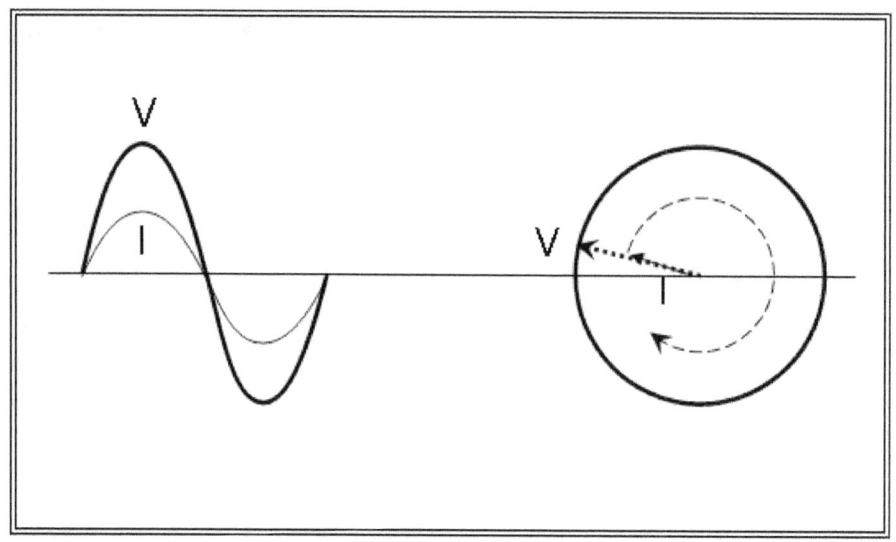

VOLTAGE AND CURRENT IN A RESISTOR
FIGURE 1-2

AC / DC

Direct Current - DC

A fresh D cell battery generates 1½ volts across its terminals. Over any period of several seconds, this remains unchanged. Connect a one ohm resistor across it, and electrons flow out of the negative, butt end, through the resistor, and into the positive, button end.

By an historic error, we say *current* flows the other way, from positive to negative, and 1½ amperes passes through the resistor. A simple equation connects the three: $V = I R$. Voltage equals current times resistance. Flipped about, it's ohm's law: $R = V / I$: resistance is the ratio of voltage divided by current.

Alternating Current - AC

In this book we are interested in voltage that varies a certain way with time, in an oscillating positive / negative pattern called a sine wave. Graphically, we show this by thinking of a spinning arrow called a *phasor*. Let me explain by example.

In Figure 1-1, an arrow spins at a constant rate about its tail. It traces out a circle. At every point on the circle, the distance between its horizontal diameter and the arrow head represents a voltage, so the curve to the left results. The horizontal axis is the angle of the arrow from horizontal. This is a *sine* wave. If the arrow spins 60 times per second, the voltage has a frequency of 60 cycles per second, or 60 Hz. We will be interested in cases where the frequency is much higher, in the kHz, MHz, GHz, and such ranges.

Resistance

When the voltage is placed across a resistor, current flows. The current is also a sine wave, and is in step with the voltage at all times: they have the same phase. Figure 1-2 shows this case. I've made the current one-half the voltage, a resistance of 2 ohms.

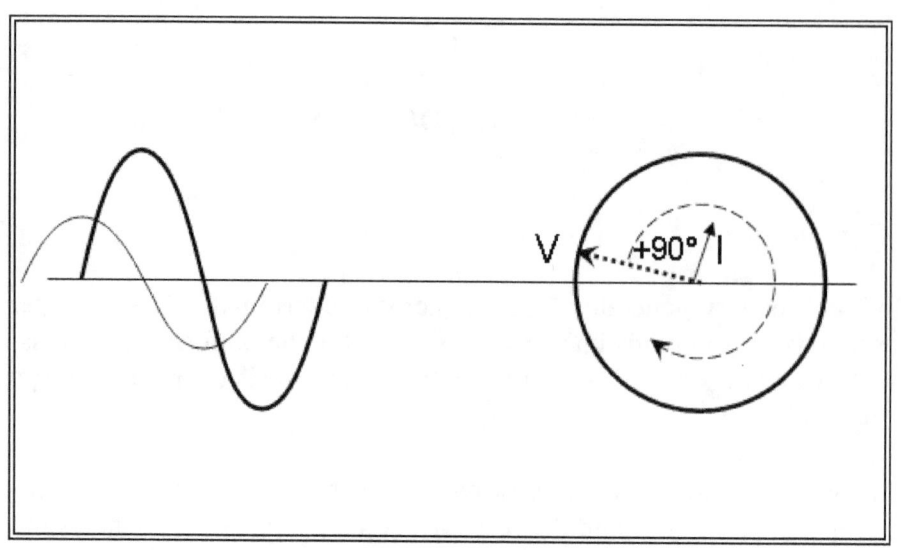

VOLTAGE LAGS CURRENT 90°
CAPACITIVE REACTANCE
FIGURE 1-3

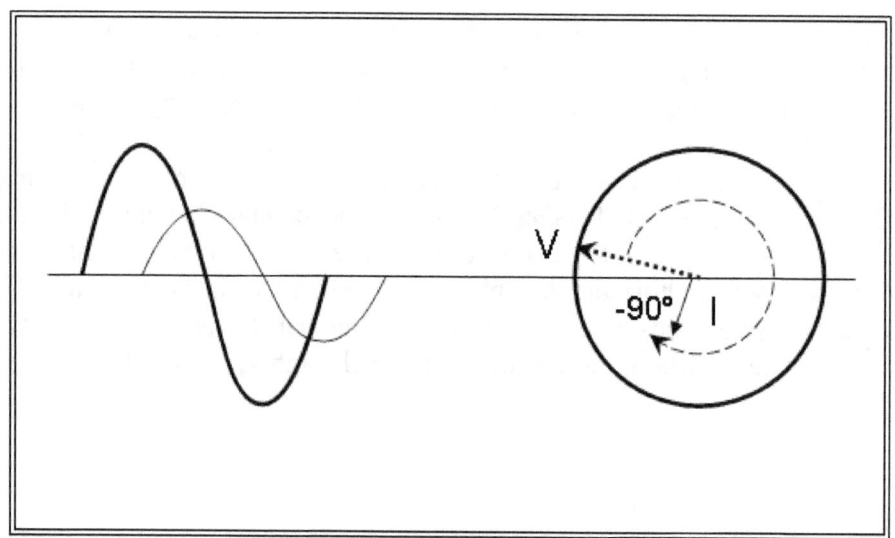

VOLTAGE LEADS CURRENT 90°
INDUCTIVE REACTANCE
FIGURE 1-4

Reactance

But often the current is *not* in phase. When the voltage is across a pure capacitor, the current depends on the rate at which the voltage changes, as in Figure 1-3. When the voltage rises most rapidly, current has its maximum positive value. When the voltage is at its maximum or minimum, current is zero, and when the voltage is dropping most rapidly, current has its maximum negative value as seen.

If the voltage is one volt peak, the current shown is one-half ampere peak. Again ohm's law holds, but the voltage lags (is later in time than) the current by a quarter cycle, or 90 degrees. One volt still produces a half ampere. This is still a two ohm situation but it is *reactance*, and capacitive reactance at that. And of course, the opposite can occur. In Figure 1-4, the voltage depends on how fast the current changes. Then the voltage leads (is earlier in time than) the current by 90 degrees. This is the case again of two ohms, inductive reactance.

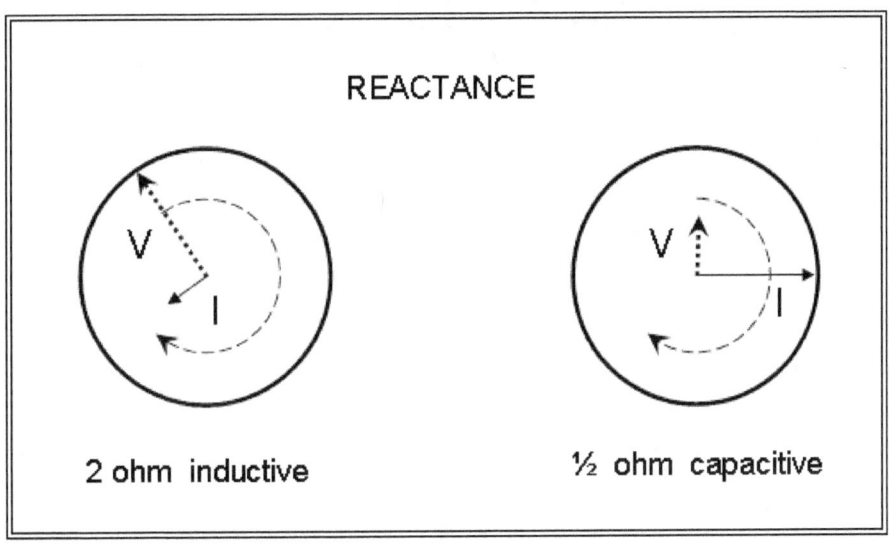

EXAMPLES OF REACTANCE
FIGURE 1-5

Now we can largely put away sine waves, using phasors instead. Let's get familiar with some examples.

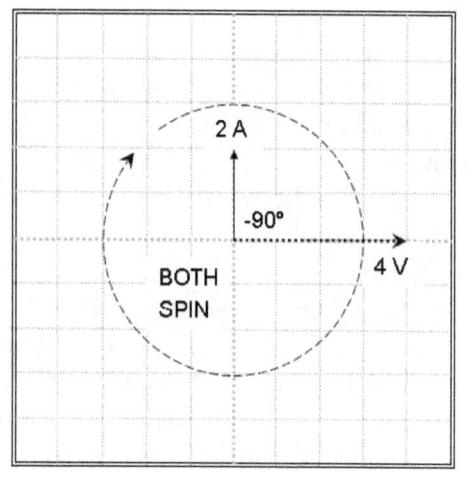
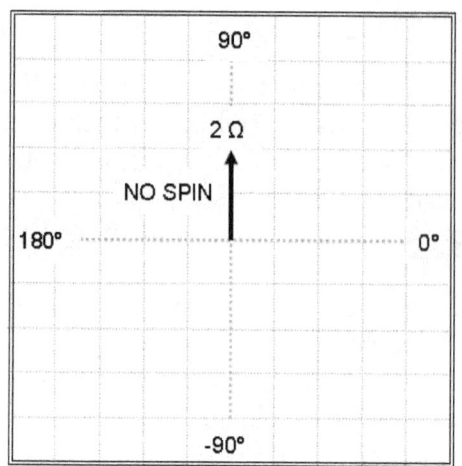

Voltage and current phasors　　　　Reactance phasor

**INDUCTIVE REACTANCE
VOLTAGE LEADS CURRENT
CURRENT LAGS BEHND VOLTAGE
FIGURE 1-6**

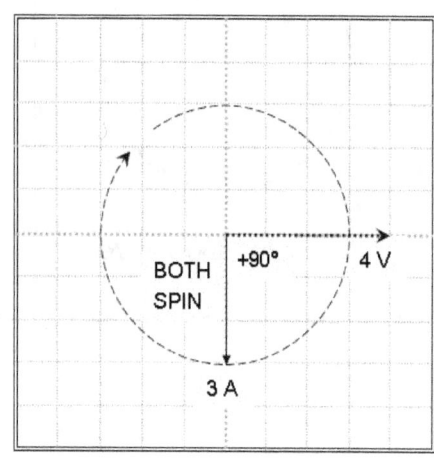
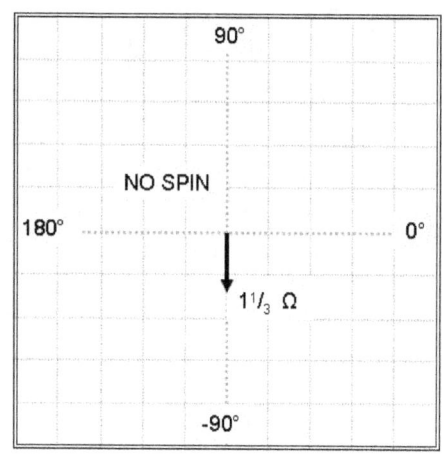

Voltage and current phasors　　　　Reactance phasor

**CAPACITIVE REACTANCE
CURRENT LEADS VOLTAGE
FIGURE 1-7**

Figure 1-6, left side, shows 4 volts producing 2 amperes, 90 degrees late. To calculate the reactance we divide voltage by current as follows:

$$V / I = 4V \angle 0° / 2A \angle -90°$$

The notation $\angle -90°$ states that the current lags behind the voltage by 90 degrees, one-quarter of a cycle. We will find other ways to express the same shortly.

Here, get the reactance by dividing the top magnitude, 4, by the bottom magnitude, 2, and find the angle by **subtracting** the bottom angle from the top angle. The result is a phasor, $2\Omega \angle +90°$. It does not spin. Get capacitive reactance of $1\frac{1}{3}\Omega \angle -90°$ by doing the same with the other example, shown in Figure 1-7.

Impedance

Of course, we can represent resistance in this same notation. Figure 1-8 shows a voltage, $4.2V \angle +45°$, a current of $1.4A \angle +45°$, a resistance, $3\Omega \angle 0°$. The chart on the right, showing resistance and reactance, is an *impedance* chart. We label the horizontal axis R, for resistance, and the vertical axis X, for reactance.

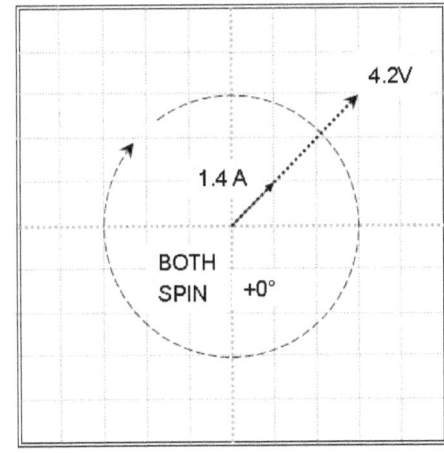
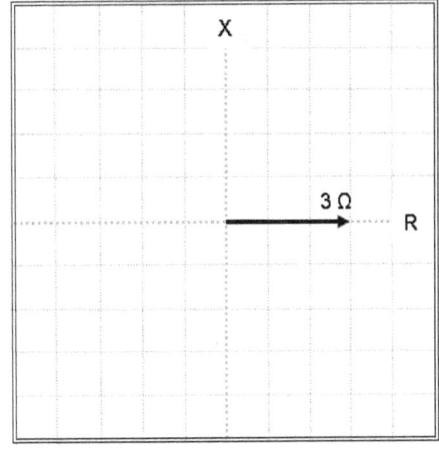

RESISTANCE ON AN IMPEDANCE CHART
FIGURE 1-8

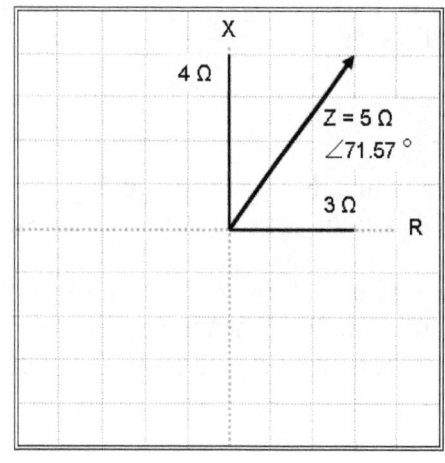 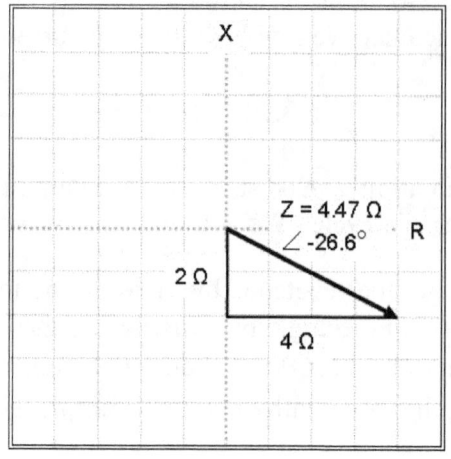

Z = 3Ω +j 4Ω   Z = 4Ω –j 2Ω

TWO IMPEDANCES
FIGURE 1-9

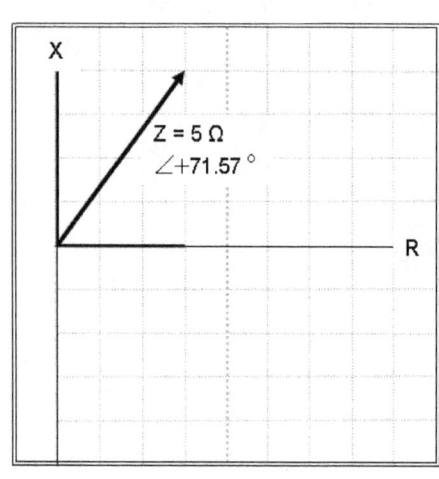 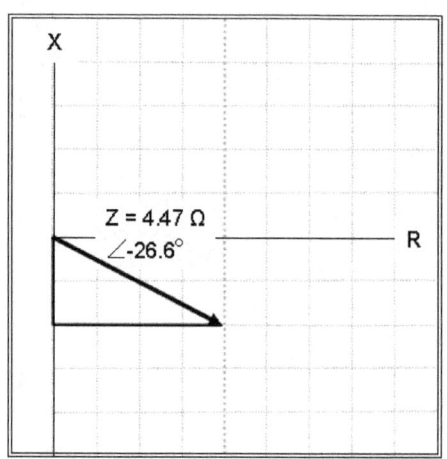

Z = 3Ω +j 4Ω   Z = 4Ω –j 2Ω

IMPEDANCE CHART
RECTANGULAR FORM
FIGURE 1-10

Figure 1-9 shows two different impedances. When resistance and reactance are combined in a series circuit, the phase angle between current and voltage can be other than zero or 90 degrees: it can lie anywhere between −90° and +90°.[1] The symbol Z is used to denote impedance, a phasor sum of resistance and reactance. And when given some thought, we realize resistance in non-amplifying systems, such as coaxial cable and other transmission lines, can never be negative. Therefore we can redo impedance charts by removing the left half: Figure 1-10 is just Figure 1-9 so modified, and is in a fairly standard form.

Notations such as 5 ∠+71.57°, seen in Figure 1-10, left, is expressible in another form. Notice that the impedance is 3 ohms resistance and 4 ohms inductive (positive) reactance. We often write Z = 3 +j 4 ohms. The "j" is simply a jog factor, telling us to turn the phasor counter-clockwise 90 degrees. In mathematics, this notation denotes a *complex number*. The resistance axis (horizontal) is called the *real* axis, and the impedance, j axis, is called the *imaginary* axis. These notations will be used rather glibly and interchangeably.

Normalized Impedance

Any antenna has a *feed* point where coax is attached. The antenna presents some impedance at this point, and the impedance varies with frequency. A typical dipole might have the following impedances:

Frequency MHz	R ohms	X ohms	Frequency MHz	r	x
14.0	52	-60	14.0	1.04	-1.20
14.1	59	-22	14.1	1.18	-0.44
14.2	70	0	14.2	1.40	0
14.3	89	26	14.3	1.78	+0.52
14.4	113	58	14.4	2.26	+1.16

ANTENNA IMPEDANCE
TABLE 1-1

I'll use 50 ohm coax to feed this antenna. It's industry standard to convert all impedances to *normalized* form: $z = Z / Z_o$, $r = R / Z_o$, and $x = X / Z_o$, where Z_o is the coax line's impedance. The table includes the normalized values, r and x. *Hint*: these are important parameters.

---

[1] It's restricted to that range because resistance cannot be negative.

The initial and normalized values of a particular antenna's impedance are plotted on an R, X rectangular graph in Figure 1-11. We'll shortly see how to plot these in more useful form – the object of this book.

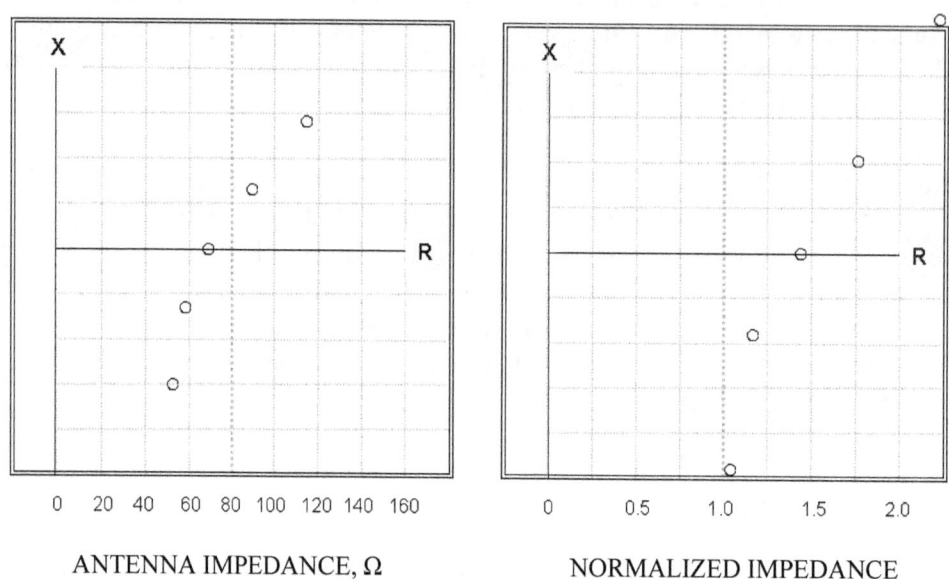

ANTENNA IMPEDANCE, Ω   NORMALIZED IMPEDANCE

IMPEDANCE PLOTS
FIGURE 1-11

COAX AND ITS IMPEDANCE

Characteristic Impedance

Get a piece of RG-8 coax cable a quarter wavelength long with a connector on one end. For example, a 33 cm piece at 150 MHz. will do.[2] Put the connector on an antenna analyzer. Leave the far end open, connected to nothing. At 150 MHz, you'll see Z = 0 +j 0 ohms: a short circuit. And by measuring it at half frequency, 75 MHz, you'll find Z = 0 –j 50 ohms, a pure capacitance. I don't go into the math here, but just make these statements.

If you find Z = 0 –j 75 ohms at 75 MHz, you've got the wrong coax. It's probably RG-11, 75 ohm line. Go get some 50 ohm line before continuing. This, incidentally, is a direct way to measure the *characteristic impedance* of the coax, and its loss. The resistive part should be very close to zero ohms. If it is higher, the coax is lossy.

Now put a 50 ohm resistor load on the open end of the coax, and see Z = 50 +j 0 ohms at 150 MHz, at 75 MHz, at 225 MHz, and at all frequencies. This is the *characteristic impedance* of the cable. Put a 100 ohm resistor load on its end, and find that the reading at 150 MHz is now Z = 25 +j 0 ohms. And if you put a 42 pF capacitor across the open end, you'll find Z = 0 +j 25 ohms (sic).

OK – I've been glib and danced around all this. Just how the heck did I come up with this stuff? Aha! That's the neat part of this book. Let's continue …

Two Constants

Well, we see that the impedance at one end of coax isn't the same as that at the other end, except in a specific case. Impedance seems to change, depending on the length of the coax. But there's a parameter that is constant. It's called the Standing Wave Ratio, SWR, also Voltage Standing Wave Ratio, or VSWR. No matter where you cut the line, the VSWR will always be the same.[3]

10

---

[2] RG-8, length ¼ x 300/150, times its velocity factor of 2/3.
[3] For lossless line. For very long, real lines, VSWR will slowly reduce along the length.

But VSWR only tells us a magnitude. We need another parameter, another that stays constant along the coax, and this is called the *reflection coefficient*, with symbol ρ. This parameter is a complex number. VSWR is related to its magnitude, |ρ|, by VSWR = (1+|ρ|) / (1−|ρ|).

The Reflection Coefficient

A coaxial line has a characteristic impedance called Zo. For RG-8, Zo = 50 ohms: for RG-6 it's 75 ohms. These values are the most common in the industry, tho some odd ball impedances exist here and there. You'll probably never meet any of them, but be aware. You already know how to measure Zo for any type of coax. We accommodate Zo by using the normalized impedance, $z = Z / Zo$.

The reflection coefficient is defined as ρ = (z−1) / (z+1). ρ is a complex number: ρ = α +j β. We'll make an x-y plot, α on the horizontal axis, and β on the vertical axis. On this plot, a circle of constant radius is one of constant magnitude of ρ, and thus one of constant VSWR.

Using the equation for ρ above, plot its real and imaginary parts, α and j β, when keeping *r* constant and letting *x* vary from −200 to +200. For 50 ohm coax, this is a range of reactance, X, from −10,000 to +10,000 ohms.. Do the same by keeping *x* constant and letting *r* vary from 0 to 20, which for 50 ohm coax, is a range of R varying from 0 to 1000 ohms. Here's the results.

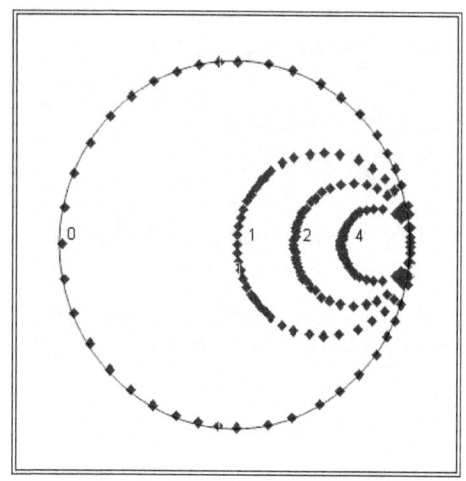 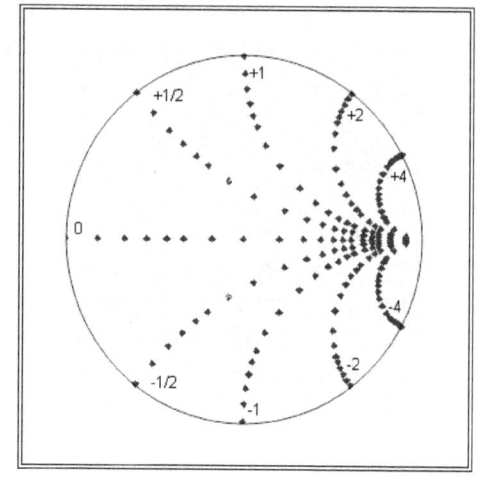

PLOT OF CONSTANT *r*   PLOT OF CONSTANT *x*

PLOT OF REFLECTION COEFFICIENT
FIGURE 2-1

All points in either plot fall inside a circle of radius one. That is, |ρ| is never greater than 1. On the left, I see four circles, for $r = 0, 1, 2$, and 4, from largest to smallest. On the right I see a straight line for $x = 0$, and four pairs of circular arcs. The top set occurs for $x = 0.5, 1, 2$, and 4, and the bottom set for the matching negative values of x. The overlap of these two plots, showing a graphic relationship of ρ to normalized impedance $r + j\,x$ is the smith chart. Its great utility is that, as one moves along a coaxial transmission line, the impedance simply runs around a circle of constant |ρ| or constant VSWR, centered at the chart's center.

There are several properties of these circles and circular arcs. First, the resistance circles are based on a simple fraction, $1/(r+1)$. Each circle has a radius proportional to this fraction. For example:

$r = 0$ Radius $= 1/(0+1) = 1$
$r = 1$ Radius $= 1/(1+1) = ½$
$r = 2$ Radius $= 1/(2+1) = ⅓$

All circles have diameters on the same horizontal line, and all are tangent at the right side of the chart.

The reactance arcs are parts of circles, with radius equal to $1/x$. For example:

$x = ½$ Radius $= 2$
$x = 1$ Radius $= 1$
$x = 2$ Radius $= ½$
$x = 4$ Radius $= ¼$

The arcs in the top half occur for positive x, or inductive reactance, and the arcs in the bottom half for negative x, capacitive reactance. And all circles are tangent to the horizontal, $x = 0$ line at the right side of the chart. With these simple rules you can draw a smith chart on a blank piece of paper. That will be our next chore.

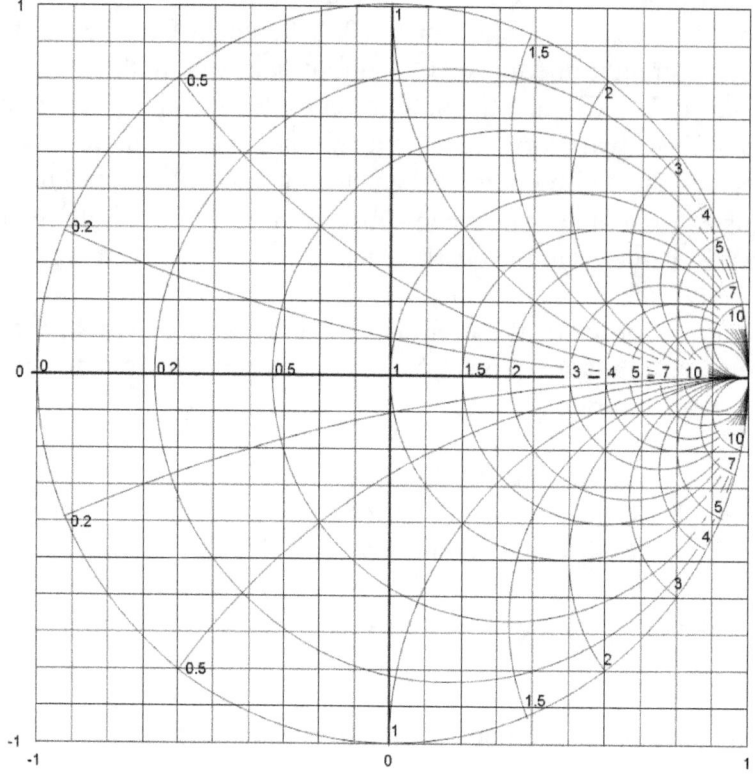

A FIRST SMITH CHART
FIGURE 3-1

The chart is simply an x,y plot of the real and imaginary parts of ρ.

THE DEFINING EQUATIONS:

$\rho = (z-1)/(z+1) \;\; \rightarrow \;\; \rho = 1 - 2(r+1)/D \;\; -j\, 2x/D$
$D = (r+1)^2 + x^2$
$\text{Re}(\rho) = \alpha = 1 - 2(r+1)/D$
$\text{Im}(\rho) = \beta = 2x/D$

$z = Z/Z_o \qquad r = R/Z_o \qquad x = X/Z_o$

While Z, R, and X are in ohms, z, *r*, and *x* are "dimensionless" parameters: they are simple ratios of two impedances.

DRAWING A SMITH CHART

The chart is just an x,y plot of the reflection coefficient, the real part on the x axis, and the imaginary part on the y axis. Figure 3-1 shows one, fairly advanced in construction. To draw it, draw a large circle and its horizontal diameter on a blank piece of paper: use about 80 percent of your paper. Label the circle "0" on the top, left-hand side of the diameter. This is the $r = 0$ circle.

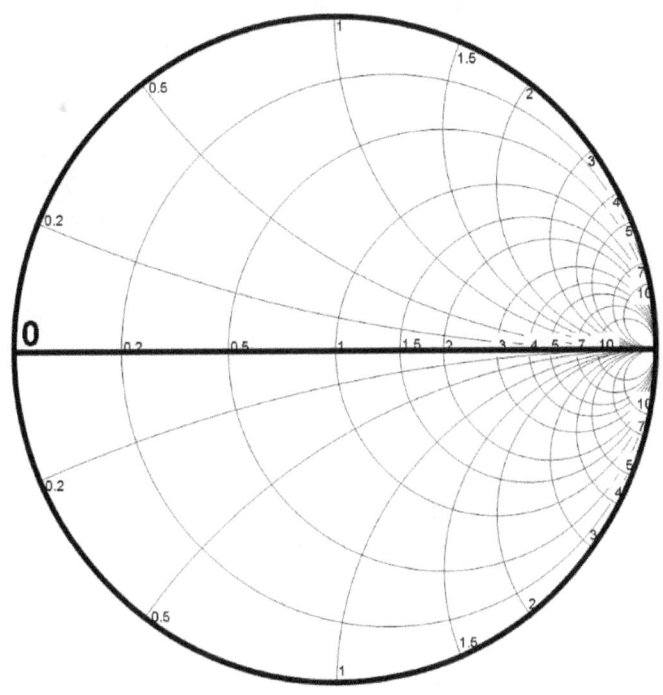

$r = 0$ CIRCLE
RADIUS IS PROPORTIONAL TO $1 / (r + 1)$
FIGURE 3-2

The radius of the circle is a scaling factor, k. If it's 8 inches, then k = 8. The radii of all other circles and parts thereof will be fractions or multiples of k.

Resistance Circles

Draw circles with radii that are fractional ratios of k, using $1/2$, $1/3$, $1/4$. Label them 1, 2, and 3. Draw all tangent at the right side of the chart.

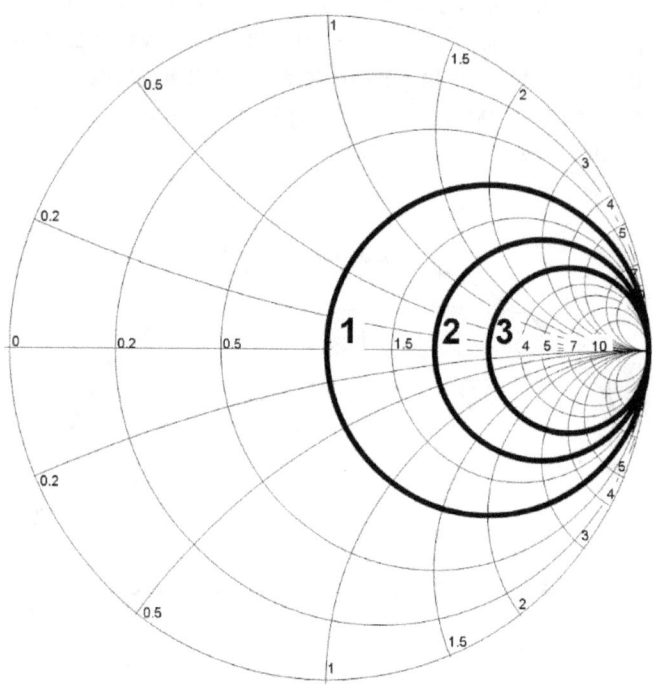

NORMALIZED RESISTANCE 1, 2, AND 3 CIRCLES
RADIUS IS PROPORTIONAL TO $1/(r+1)$
FIGURE 3-3

Each circle plots a normalized resistance, r. For each, the inverse of its radius, minus one, is equal to r. Thus, for 1, it's zero. For $½$, it's 1. For $1/3$, it's 2. For $¼$, it's 3. And for $2/3$, it's $½$. For $10/12$, it's $2/10$. Work it out to see this.

To draw other resistance circles, for example, $r = 0.25$, add one and invert the result to get the required radius. i.e, $0.25 + 1 = 1/4 + 1 = 5/4$. Radius of this circle is $4/5$ of the outer circle radius. Complete the resistance circles this way. You will now have all the circles shown in Figure 3-1. You can add as many more as you may desire. Since they get bunched up at the right side of the chart, it's often drawn with some circles partially completed, as circular arcs.

Reactance Arcs

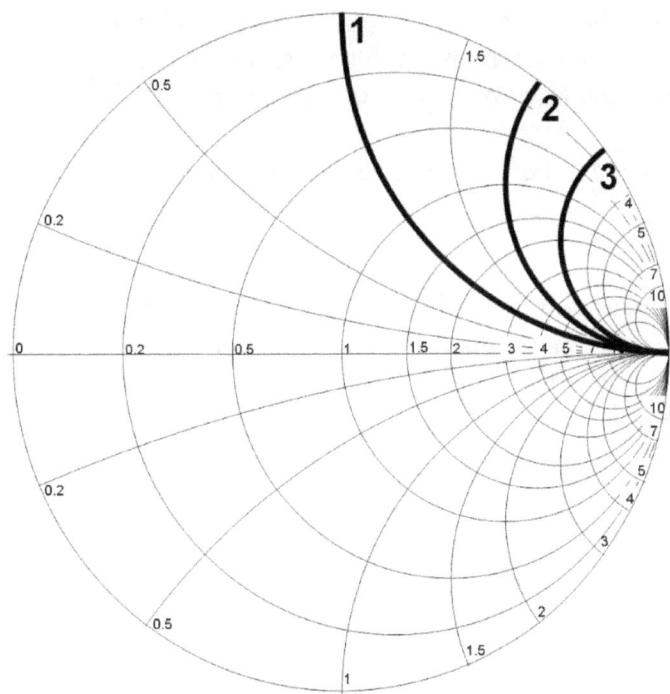

**NORMALIZED REACTANCE 1, 2, AND 3 ARCS
RADIUS IS PROPORTIONAL TO 1 / x
FIGURE 3-4**

Now let's tackle the reactance arcs. These are parts of circles with radius proportional to $1/x$, all tangent to the horizontal line at the right side of the chart. To get any arc, calculate its radius as the inverse of x. Draw the $x=1$, $x=2$, and $x=3$ arcs. Their radii are 1, $\frac{1}{2}$, and $\frac{1}{3}$. Label them 1, 2, and 3. Then continue to fill in all the other arcs above the horizontal line, and make mirror images of all below the horizontal line. The top arcs are for positive, inductive reactance, and the bottom set, for negative, capacitive reactance.

Here, the radius can be greater than one. For $x = 0.5$, its 2, and for $x = 0.2$, it's 5. Fill in all the arcs you think are necessary. One "arc," of infinite radius, a straight line, lies on the horizontal diameter. Its reactance is zero: this line is therefore called the real axis.

Draw as many additional reactance circles as you wish. For example, if $x = 1.5 = \frac{3}{2}$, its arc has radius $\frac{2}{3}$ of the outer circle radius.

The Chart

By choosing the set I have done, you will draw Figure 3-1, minus the rectangular grid lines. The latter are not shown on smith charts. I do so only to remind you that we are plotting the real and imaginary parts of reflection coefficient.

Since ρ is a complex number, it can be represented in magnitude, angle format.

$$\rho = \alpha + j\beta = |\rho| \angle \theta$$

The chart is customarily drawn with an angle scale around its periphery. I've added a generic angle scale about the chart in Figure 3-5 below, and have placed no numbers about the circumference, for a good reason.

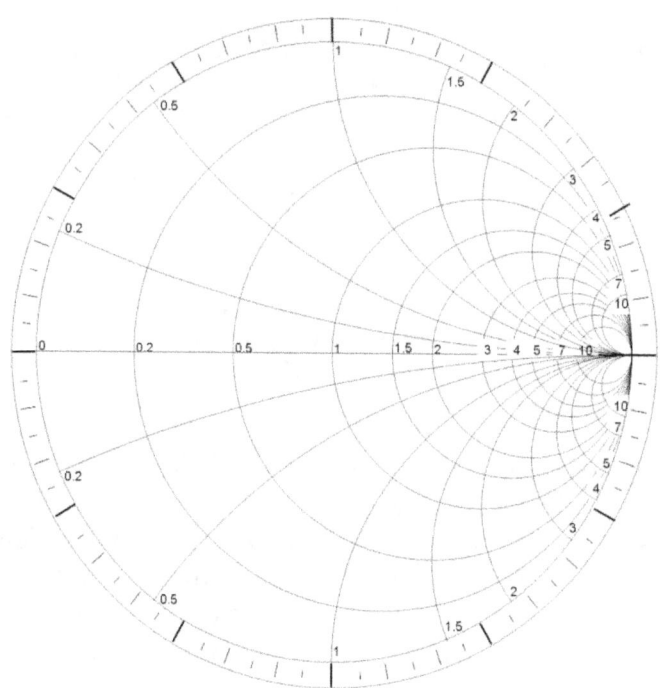

AN ANGLE SCALE, IN FIVE DEGREE STEPS
FIGURE 3-5

One scale, beginning with zero degrees at the right hand side, continues counter-clockwise with 90° at the top, 180° at the left side, 270° at the bottom, to 360° back at the right side, and is the phase angle, θ, of $\rho = |\rho| \angle \theta$. This is not often shown on the chart and of little practical use. But see Chapter 10 for exceptions.

With $\rho = (z - 1) / (z + 1)$:

For z = infinite	$\rho = +1$	$\theta = 0°$
For $z = +j$	$\rho = +j$	$\theta = 90°$
For $z = 0$	$\rho = -1$	$\theta = 180°$
For $z = -j$	$\rho = -j$	$\theta = -90°$

This is the basic form of a smith chart. All are intersecting circles and circular arcs, the former for specific values of normalized resistance, and the latter for specific values of normalized reactance.

The angle scale allows one to plot the phase angle of ρ. Its magnitude, $|\rho|$ is the distance out from the center. Constant magnitude defines a circle centered on the chart's center. This circle will be key to working with the chart in later instruction.

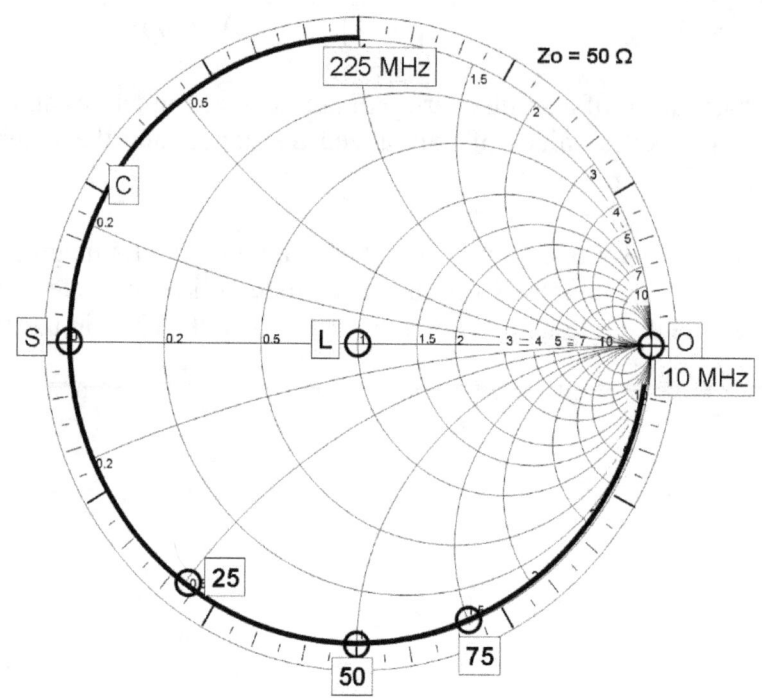

AN OPEN-CIRCUITED LINE
MEASUREMENT OF Zo AND VELOCITY FACTOR
FIGURE 4-1

FAMILIARIZATION

An Impedance Meter

Now let's get familiar with the smith chart. We'll use a virtual *Vector Network Analyzer*, or VNA, an instrument that measures RF impedance and displays results on a smith chart, right on its screen.[4] Set it up to measure the 33-cm piece ($^1/_3$ meter) of RG-8 mentioned in chapter 2. At $300 / (^1/_3) = 900$ MHz, its one wavelength long, and a quarter wavelength at $900 / 4 = 225$ MHz. Therefore, we set the VNA to scan from its low limit, say, 10 MHz, to 225 MHz.

VNAs are commonly designed to work with Zo = 50 ohms. They come with a *calibration tee*, a three-coax fitting assembly in a T shape. One arm provides a calibrated open circuit, the second, a calibrated short circuit, and the third, a calibrated 50 ohm load. With the *open* on the meter, we see a measurement on the chart at the right hand side, marked as **O** on Figure 4-1. Connecting its *short* moves the measurement to the left hand side, marked as **S**. Connecting the *load* moves the measurement to the center of the chart, marked **L**.

Impedance Measurement

Connect one end of the coax in place of the tee. Curve **C** appears on the chart as the VNA sweeps from 10 to 225 MHz, 10 MHz near point **O**, 75 MHz at the bottom, 150 MHz at point **S**, and 225 MHz on the top. At 150 MHz, the open-circuit impedance at the other end of the coax is transformed to $z = 0 +j\ 0$, a short circuit. The angle about the smith chart is 180°, but the coax is a quarter wavelength, 90 degrees, long The angle scale is then read so one complete circle represents 180° along the coax.

At 75 MHz the curve is at point **50**, an impedance of $z = 0 -j\ 1$ ohms. The coax has Zo = 50 ohms. If instead 75 MHz produces point **75** where $z = 0 -j\ 1.5$, you have 75 ohm coax, perhaps RG-11. Go back and get the right stuff.

20

---

[4] The same procedure can be done with an antenna analyzer, and making notations of its readings vs. frequency on a paper copy of a smith chart.

Velocity Factor

A quarter wavelength in air occurs at 225 MHz. In the coax, it occurs at a lower frequency, 150 MHz, because the radio wave travels slower in coax than in air. The ratio, 150 / 225 = 0.667, is called the *velocity factor* of the coax.

This simple measurement has allowed measurement of the Zo and velocity factor of a sample of coax cable. The magnitude of ρ stays constant, at one, and the VSWR remains infinite in this example.

Sense of Rotation

Consult Figure 4-2 below. A short wire connected in place of the coax produces a short curve, from the left end of the chart, proceeding clockwise a bit with increase in frequency. This is an inductor. The reactance is inductive. If a small capacitor is connected, a short curve starting at the right end of the chart proceeds clockwise a bit. The reactance is capacitive. The same curves occur if a short piece of coax is connected that is shorted at its end, and open, respectively. Loads on coax have curves proceeding from the load, clockwise, to the source of RF.

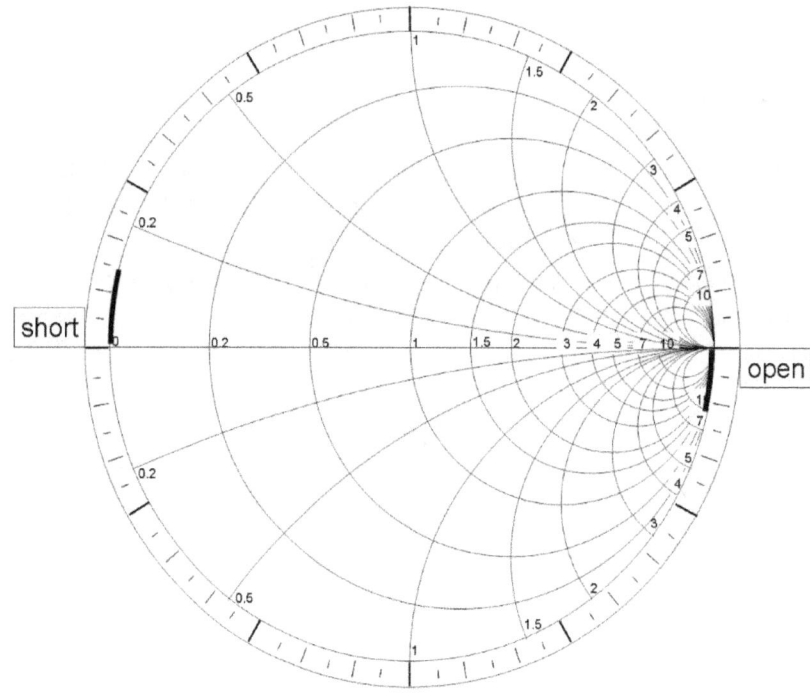

SENSE OF ROTATION
FIGURE 4-2

Plotting Normalized Impedances

Let's go back to Table 1-1, a list of normalized impedances for a particular antenna, and plot them on the chart. These are tabulated on Table 1-2, reproduced below for convenience.

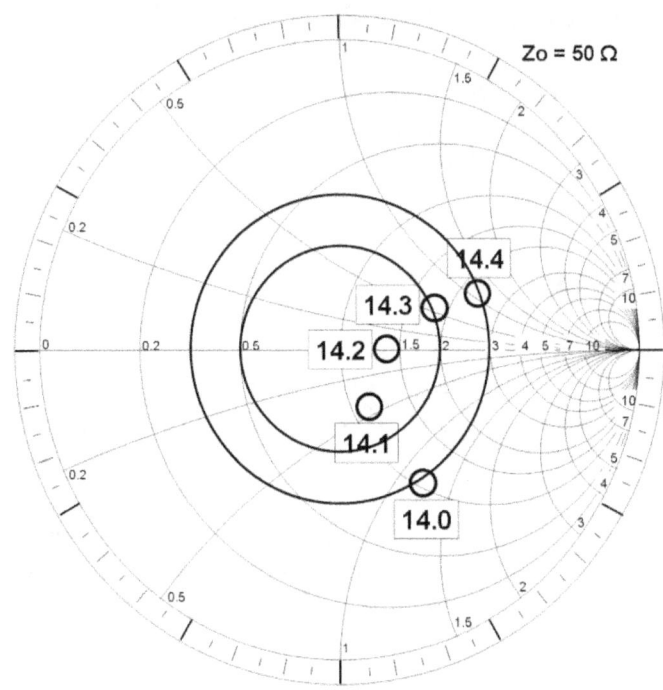

NORMALIZED ANTENNA IMPEDANCE
FIGURE 4-3

Frequency MHz	r	x
14.0	1.04	-1.20
14.1	1.18	-0.44
14.2	1.40	0
14.3	1.78	+0.52
14.4	2.26	+1.16

TABLE 1-2
NORMALIZED IMPEDANCES

Figure 4-3 (previous page) shows the result. I've introduced another set of circles on the plot. The smaller, passing through $r = 2$, is a circle with $|\rho|$ constant, and is $1/3$ the diameter of the outer circle. Thus $|\rho| = 1/3$. The larger, for $|\rho| = 1/2$, passes through $r = 3$. From earlier comment, recall that the magnitude of ρ and VSWR are related: VSWR = $(1+|\rho|) / (1-|\rho|)$. The smaller circle is the limit of all impedances with VSWR equal to 2:1, and the larger, VSWR equal to 3:1. Conveniently, these are the values of r where the circle crosses the real axis on the right side.

With this antenna, then, the VSWR is slightly over 3:1 at 14 MHz, less than 3:1 elsewhere, under 2:1 at 14.1 MHz, and about 1.5:1 at 14.2 MHz. Can the VSWR be improved? Of course. But we'll leave that to a later discussion.

Impedance / Admittance

Connect a 3 ohm and 5 ohm resistor in series. When one ampere flows through one, it flows through the other as well, with voltage drops of 3 and 5 volts, respectively, a total of 8 volts. The net resistance, V / I, is 8 / 1 = 8 ohms, the sum of 3 ohms and 5 ohms.

Now connect them in parallel. The same current does not flow through each. Instead, the same voltage appears across them. To work this out, we use a parameter called conductance, which is just 1 / resistance. The 3 ohm resistor has a conductance of $1/3$ mho,[5] and the other, $1/5$ mho. Mho is an old term you may find occasionally. Today we use the unit *siemen* instead of mho.

With 8 volts across them, the 3 ohm current is V / I = $8/3$ amperes, and the 5 ohm current is $8/5$ amperes. The total is the sum: I = $8/3 + 8/5 = 8 (1/3 + 1/5)$ amperes, or $64/15$ amperes. The net resistance is V / I = 8 / ($64/15$) ohms, or $15/8$ ohms, a conductance of $8/15$ siemens. This is just the sum of the two conductances.

The same thing happens with impedances. An impedance of 10 ohms has an *admittance*, the reciprocal of 10 ohms, of $1/10$ siemen. The inverse of impedance is called admittance, symbol Y. It's similar to conductance, but is a complex number. Look what happens if we connect an impedance of 5 +j 25 ohms in series with one of 5 –j 25 ohms. The two reactances cancel (the circuit is resonant) and we get 10 ohms.

---

[5] mho = ohm spelt backwards. The modern unit is the *siemen*.

The Rule

When two or more impedances are connected in series, we add the individual impedances: $Z = Z_1 + Z_2$. When two or more are connected in parallel, we add the individual admittances: $Y = Y_1 + Y_2$ and get the total impedance $Z = 1 / Y$.

By a gift of the gods it would seem, any impedance Z defines a specific reflection coefficient, $\rho = |\rho| \angle \theta$, and its admittance $Y = 1 / Z$ defines $\rho = |\rho| \angle(\theta - 180°)$. The magnitude remains constant, but the phase angle shifts 180 degrees.

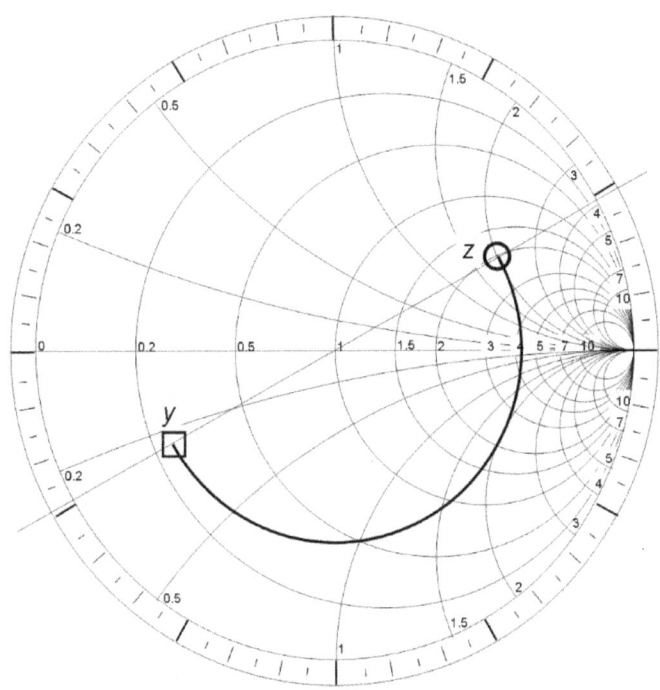

IMPEDANCE / ADMITTANCE
FIGURE 4-4

In Figure 4-4, we find that an impedance $z = 2 + j\,2$ has a corresponding admittance of $y = 0.25 - j\,0.25$. On the chart, rotate z, marked by the small circle, half way 'round the angle scale at constant radius, and mark that point with a small square. I use small circles to denote impedance, and squares for admittance, to avoid confusion. Other books and articles use differing notations.

NOTES

IMPEDANCE TRANSFORMER
RESISTIVE LOADS

The Q Section

Rotation of 180° about the chart at constant radius is the same as a move along a coax line of 90 electrical degrees. Thus a line a quarter wavelength long will move an impedance half way 'round the chart. It remains an impedance at that point by action of the line. When Z is a pure resistance, the result is that $Z_2 = Z_o^2 / Z_1$ or $Z_1 Z_2 = Z_o^2$. For example, a 50 ohm line will convert 100 ohms to 25 ohms.

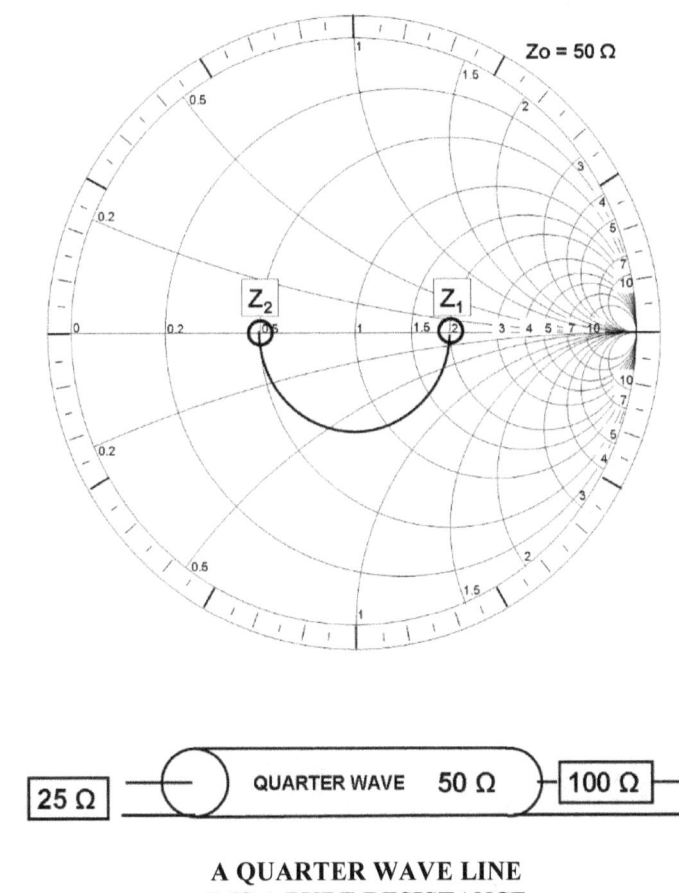

A QUARTER WAVE LINE
Z IS A PURE RESISTANCE
FIGURE 5-1

The 60 degree Balun

A half-wave dipole's impedance in free space is close to 75 ohms. I want to feed it with 50 ohm coax. To do so I start with 50 Ω ohm coax. On Figure 5-2, the dipole impedance, point **A**, has normalized impedance $z = 1.5 + j\,0$. VSWR is 1.5:1.

I cut the coax to a length of 30 electrical degrees. z is transformed to point **B**, where it is now $z = 1.2 - j\,0.4$. I switch there to 75 ohm coax. Impedance z on the new coax line becomes $2/3$ its value. $z = 0.8 - j\,0.3$, plotted at point **C**.

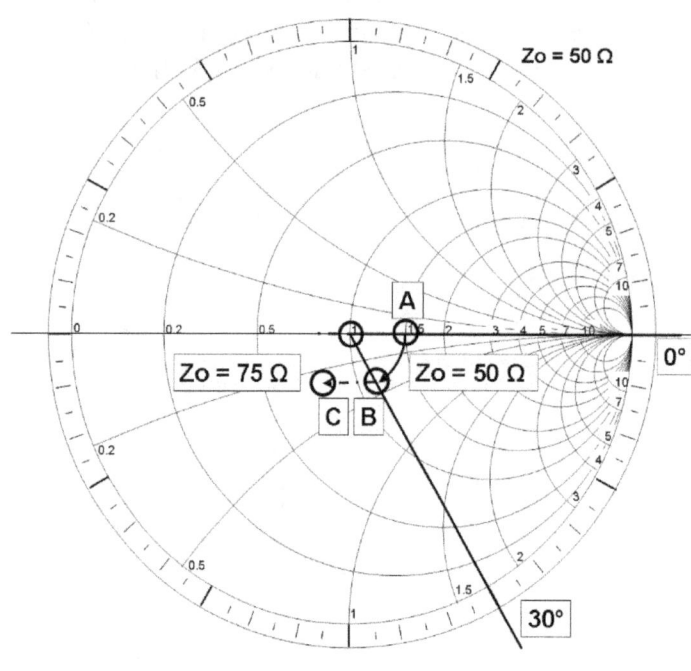

**60 DEGREE BALUN
STEP 1
FIGURE 5-2**

Go to Figure 5-3 (next page). I cut the 75 ohm line to 30 electrical degrees, and z now transforms to point **D**: $z = 0.7 + j\,0$. Then I switch back to 50 ohm coax. z is multiplied by $3/2$, and appears at point **E**: $z = 1 + j\,0$. I can now run 50 ohm coax all the way to the radio gear and it operates at VSWR 1:1. This balun has converted 75 ohms to 50 ohms.

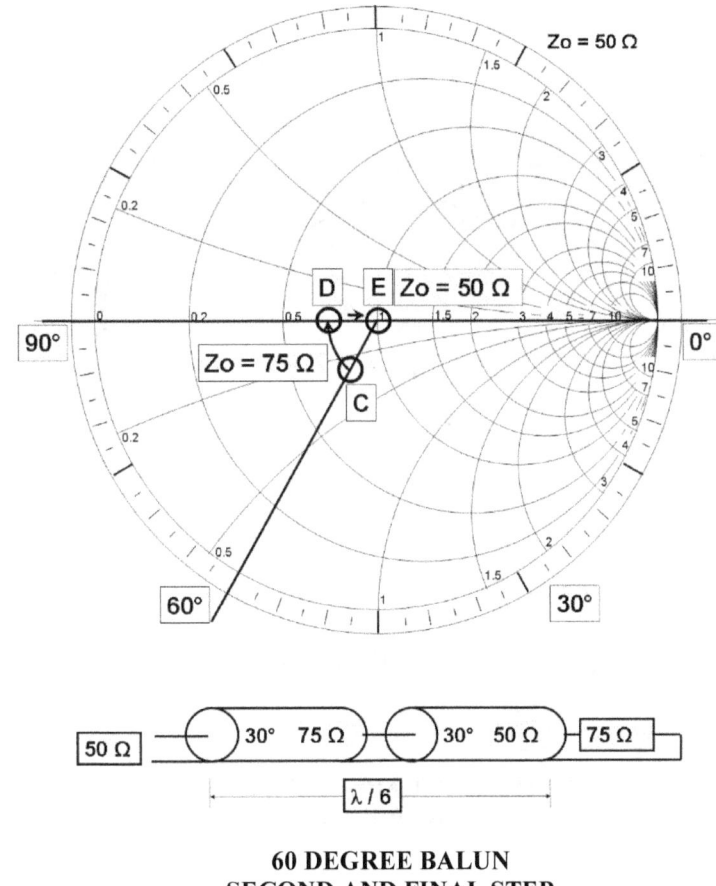

**60 DEGREE BALUN
SECOND AND FINAL STEP
FIGURE 5-3**

The same impedance transformation can be had with a Q-section of 61 ohm coax. But this stuff is not readily available. The 60 degree balun is ²/₃ the size of a Q section, and made of readily available standard coax.

A Trick – A Variable Inductor

In Chapter 2 a 42 pF capacitor was placed across the open end of the coax. It has a reactance, $X = -1/2\pi fC$ ohms, or $-j\ 25\ \Omega$ at 150 MHz. With 50 ohm coax this is a normalized impedance of $0 -j\ 0.5$ on the chart, and appears at position **42 pF** on Figure 5-4 (next page). A line 112° long transforms the impedance to $0 +j\ 10$, at position **500 Ω**. This is inductive reactance.

27

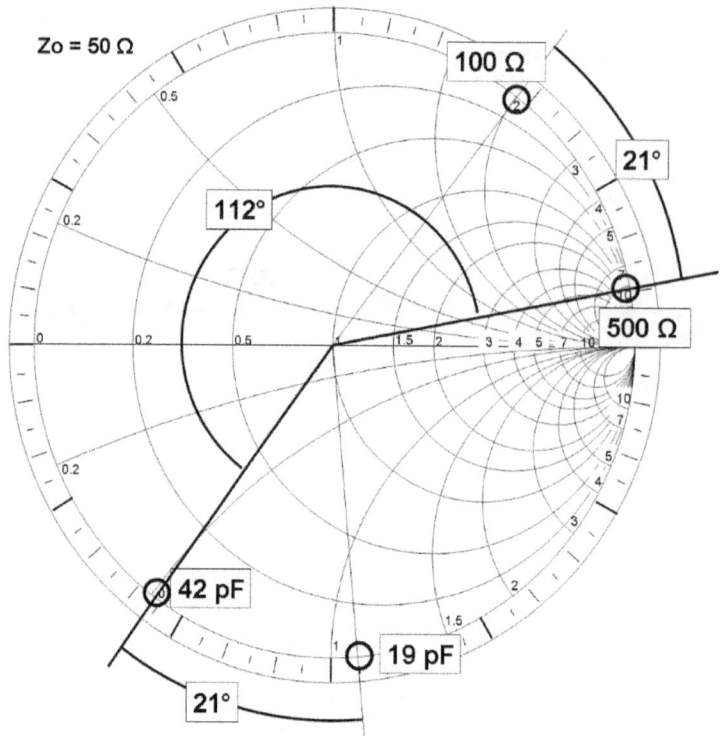

CANDIDATE VARIABLE INDUCTANCE
FIGURE 5-4

Now vary the capacitor to 19 pF, it's reactance becomes $-j\,1.1$ time 50 ohms, or 55 ohms negative reactance. This shifts its position on the chart 21° counter-clockwise. The coax transforms the impedance to $0 + j\,2.0$, or 100 ohms. This could synthesize a fairly high quality variable inductor, but the line operates at a high VSWR so will have high circulating current and possibly high voltages on it. As the Q factor of capacitors and good coax is higher than that for inductors, this could be a way to synthesize high-Q inductance.

By a small change in line length, I might effect a variable inductance from some low value up to infinite impedance: i.e., one with tremendous range. Although at this writing the concept is only one in though, it suggests an interesting way to tune a short wire antenna to 1.8 MHz or even lower. Trying it in actual practice is a project I have yet to tackle.

NOTES

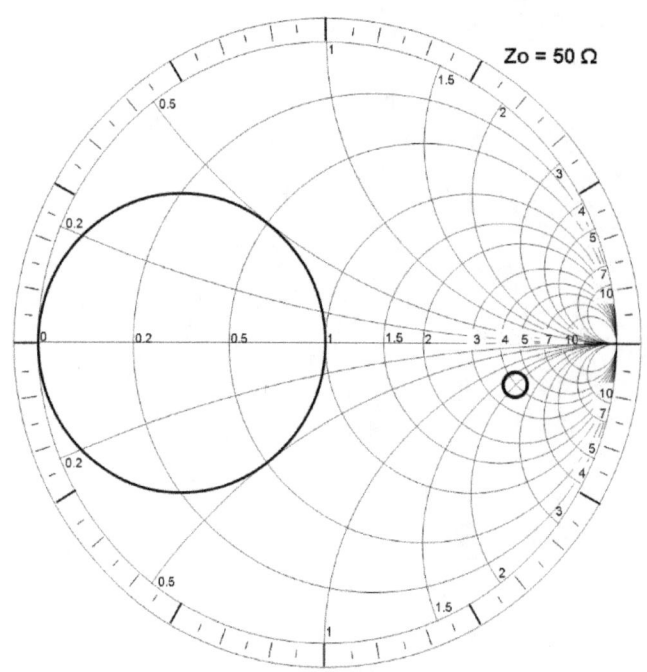

UNIT CONDUCTANCE CIRCLE
FIGURE 6-1

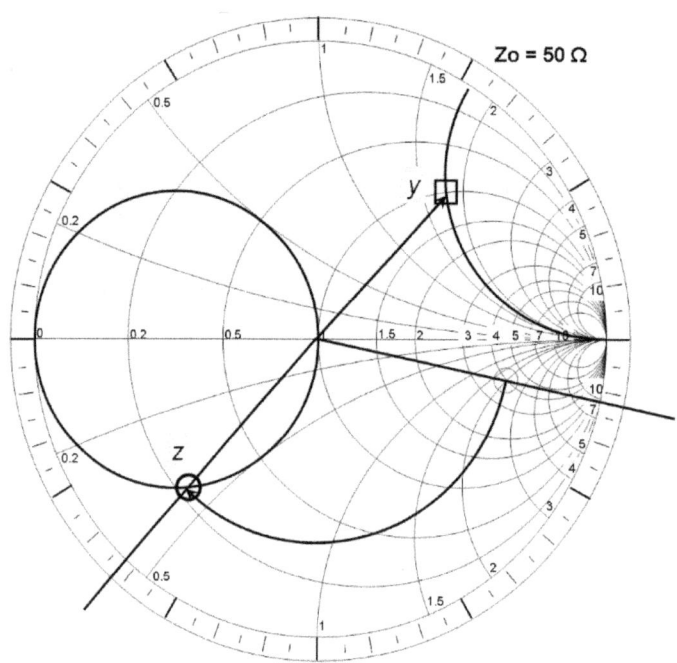

FIGURE 6-2

- 6 -

IMPEDANCE TRANSFORMER
GENERAL LOADS:
ANTENNA TUNING

An antenna's impedance is rarely resistive. At desirable frequencies it will present some reactance. On the other hand, it's industry practice to work with an *interface control standard* when designing transmitters and receivers. That is, the equipment designer often works diligently to couple his device to a 50 ohm transmission line. The antenna designer then does the same. And the coax designer concocts 50 ohm coax. This chapter examines some concepts of how to "match" the antenna to the coax impedance.

Series / Shunt Line

If I feed an antenna with 50 ohm coax[6] I can place a Tee in the line at a judiciously chosen point, and connect a short stub of coax to the line at this point. This technique will match *any* antenna that presents *any* impedance. From the Tee toward the radio gear, the VSWR can be held to well under 1:3:1. Let's see how.

On Figure 6-1 plot the antenna's normalized impedance. To demonstrate, I've chosen $z = 4 - j\,2$, an antenna exhibiting 200 ohms resistance and 100 ohms capacitive reactance. On the chart, copy the $r = 1$ circle, and move it tangent to the left hand side as shown. Any impedance on the $r = 1$ circle can be given as an admittance on this new, $g = 1$ *unit conductance* circle.[7]

Next, move along the coax at constant VSWR until reaching the unit conductance circle as in Figure 6-2. The move is 62 degrees at a VSWR of 5:1. Convert this to admittance by continuing around the chart another half turn, to the small square marked at $y = 1.0 + j\,1.7$, on the susceptance curve $b = +1.7$. This is capacitive susceptance, and can be eliminated with an inductive susceptance of equal value but opposite sign. That step is done in Figure 6-3 (next page). A piece of coax 28° long, shorted at its end, does the job. The dimensions "30°" and "60°" just came out to those values in this example. They will not be the same in general.

30

---

[6] Or transmission line of any other impedance.
[7] Just as R, resistance, is in ohms, G = 1 / R is in siemens, and is called conductance.

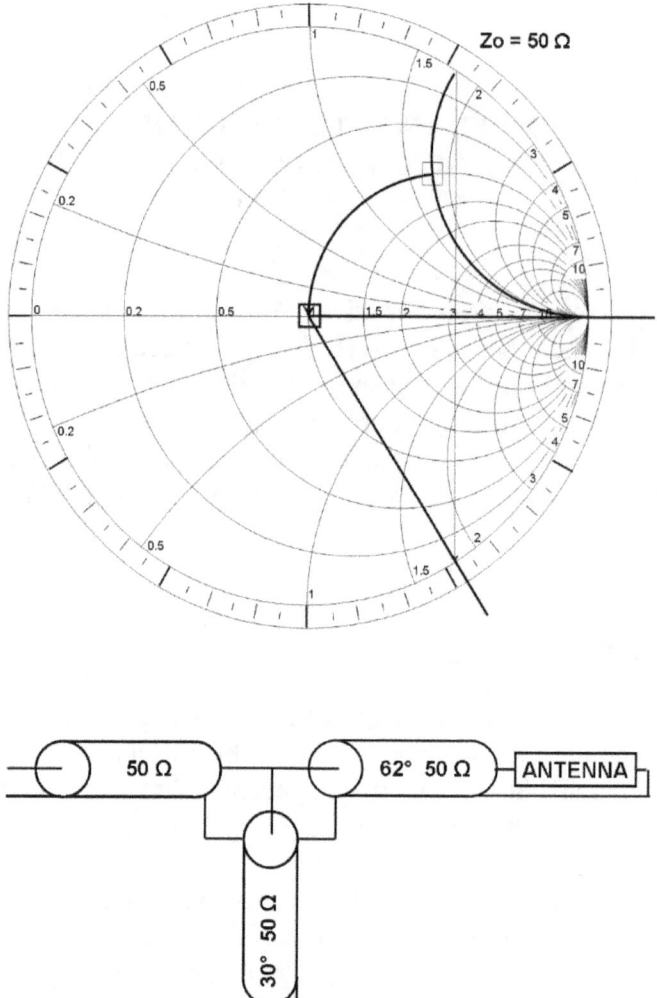

SERIES / SHUNT LINE TUNER
FIGURE 6-3

Give it some thought. You will realize that this technique can be seen as capable of working with *any* impedance, and transmission lines of any customary Zo. I'll explore it later as the means to design and match J-poles, Zepps, and some other antennas.

This technique can be reversed. If the two lines can be varied in length, and the Tee position brought to a 1:1 VSWR, then mere measurement of line lengths and use of the smith chart can reveal the antenna's feedpoint impedance. The concept is sketched in Figure 6-4, using open-wire 200 ohm line.

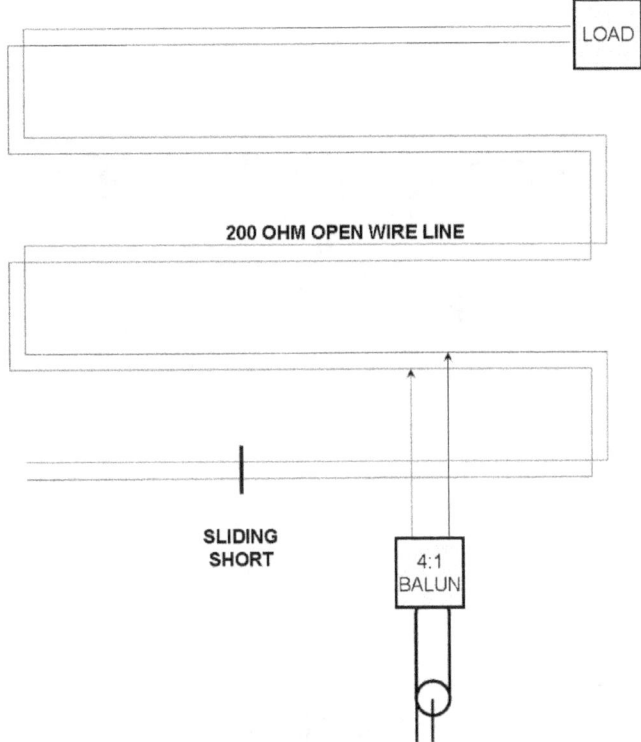

OPEN WIRE LINE IMPEDANCE MEASUREMENT
FIGURE 6-4

The open wire transmission line can be built in a few versions. One is on an outdoor clothesline rack, the other, on a sheet of plywood. The former is overhead about seven feet, the latter leaning against the house or garage external wall. The 50 ohm coax and 4:1 balun have a flexible pigtail to reach any point on the line. Its attachment point and the position of the sliding short are adjusted for 1:1 VSWR and the lengths of line from load to balun, and balun to short are the data to be recorded.

The L Network

Antennas are often tuned with L-C networks. The simplest is an L-network, an L-shaped combination of a series coil and a shunt capacitor, or a series capacitor and a shunt coil. The former is a low pass filter, and can attenuate harmonics, while the latter is a high pass filter and can pass harmonics to the antenna. Let's look at the first option.

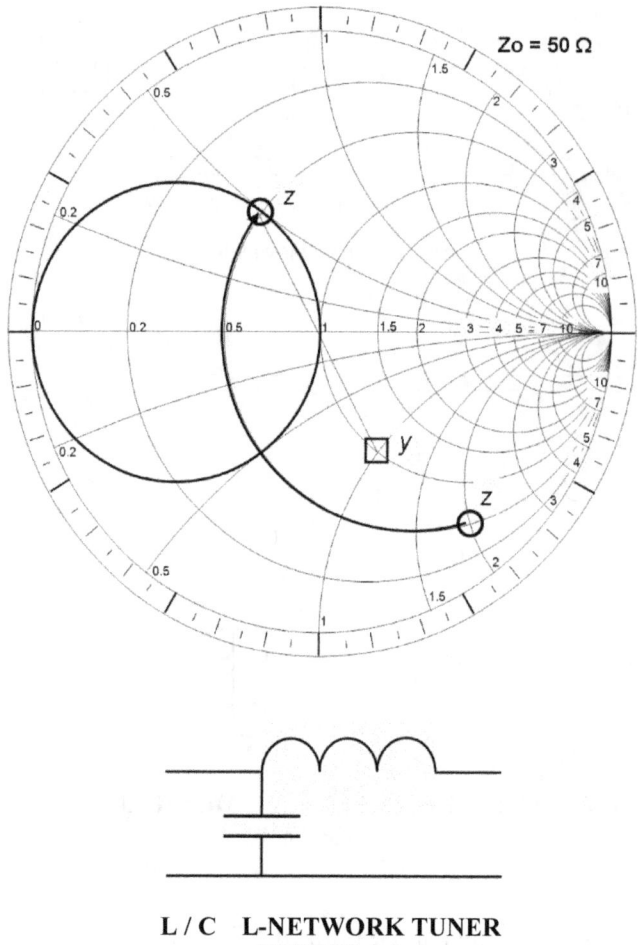

**L / C   L-NETWORK TUNER
FIGURE 6-5**

In the example the antenna impedance, $Z = 25 -j\,100$ ohms at the right side of the network, produces $z = 0.5 -j\,2.0$ The series inductor adds reactance $x = +2.5$, bringing the impedance at its left end to $z = 0.5 +j\,0.5$. Notice that the displacement is now along a line of constant r, not a constant VSWR. The equivalent admittance is found to be $y = 1.0 -j\,1.0$. The shut capacitor on the left side then provides susceptance $b = +1.0$. This brings the admittance along the line of constant conductance right to the middle of the chart, with a VSWR of 1:1.

Curiously Figure 6-5 suggests another option, as in Figure 6-6 (next page). Suppose the inductor provided only $x = +1.5$ The new impedance, $z = 0.5 -j\,0.5$, would provide an admittance of $y = 1.0 +j\,1.0$. An inductive susceptance $b = -1.0$ would bring the admittance to the center of the chart, at VSWR = 1:1. This is an L / L tuner, as in Figure 6-6, and can be useful for low resistance, high-capacity antennas such as short whip antennas.

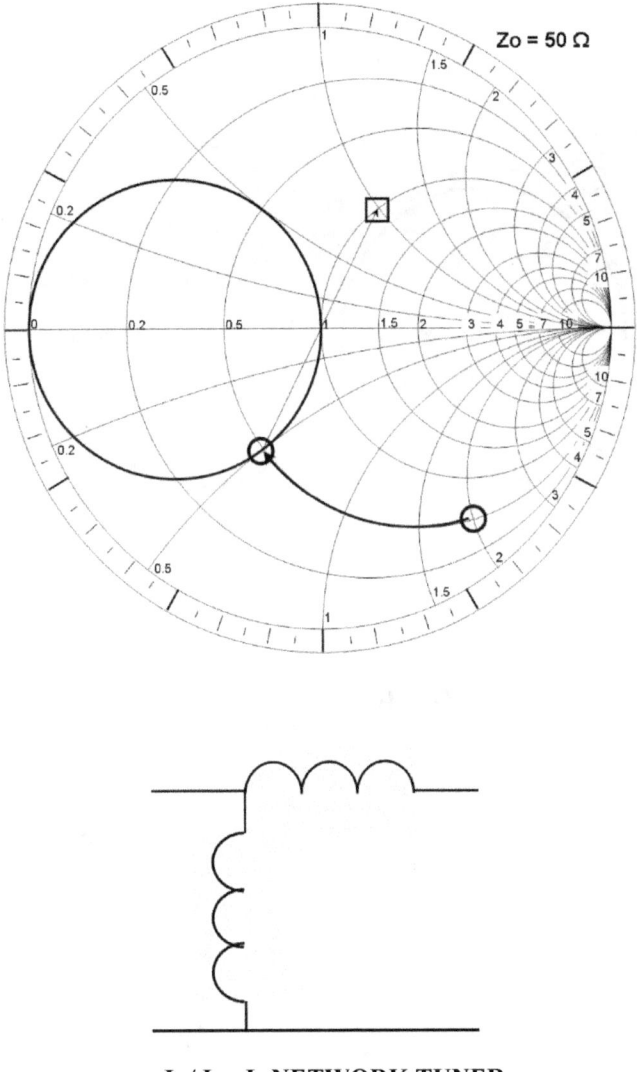

L / L   L-NETWORK TUNER
FIGURE 6-6

A Problem and its Solution

Reflect a bit on Figure 6-2. For the L-network to do its job, the antenna impedance must transform to the unit conductance circle by action of inductive reactance. This is sometimes impossible. Figure 6-7 (next page) shows a "forbidden" region, actually one-half of the chart area: Impedances in this region cannot be matched by L-networks where the antenna is fed by a series inductor with no shunt susceptance across its terminals. But we can exchange the two ports of the L-network, to provide that shunt.

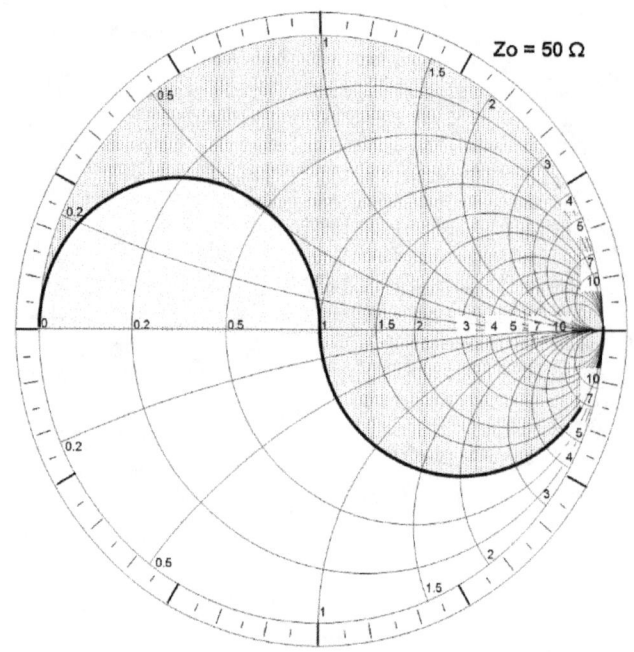

FORBIDDEN REGION
FIGURE 6-7

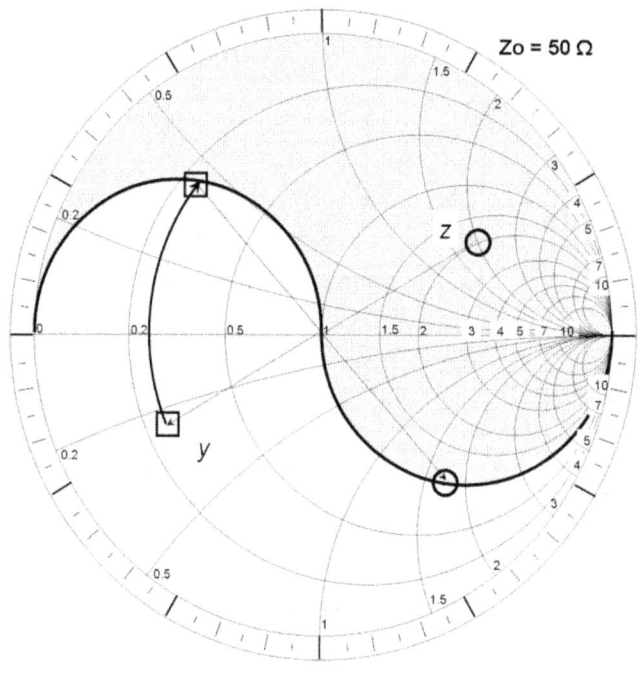

FIGURE 6-8

In Figure 6-8 an impedance of $z = 2 + j\,2$ falls in the forbidden zone. Half way 'round, the chart, its admittance is $y = 0.25 - j\,0.25$, outside the zone. A *shunt* susceptance of $b = +0.65$ brings the admittance to the unit conductance circle at $y = 0.25 + j\,0.4$, whose inverse is an impedance of $z = 1.0 - j\,1.7$. Finally, a series inductance with reactance $x = +1.7$ will bring the impedance to chart center, with VSWR of 1:1.

In this example there is only one point of contact with the unit conductance circle. A network with shunt C, series C is not useful, tho for certain cases it can be. With the knowledge you've gained so far, you can determine when: basically, z must be outside the *unit resistance* circle and in the forbidden zone: antennas with low resistance, low inductance, such as small loops are such.

Figure 6-9 shows the two zones and the type of L-network that works in each. I do not analyze the pi or tee networks here, as this is not a discussion of antenna tuners, but rather one of the smith chart and how to use it. The same principles apply: have fun as you dig into it!

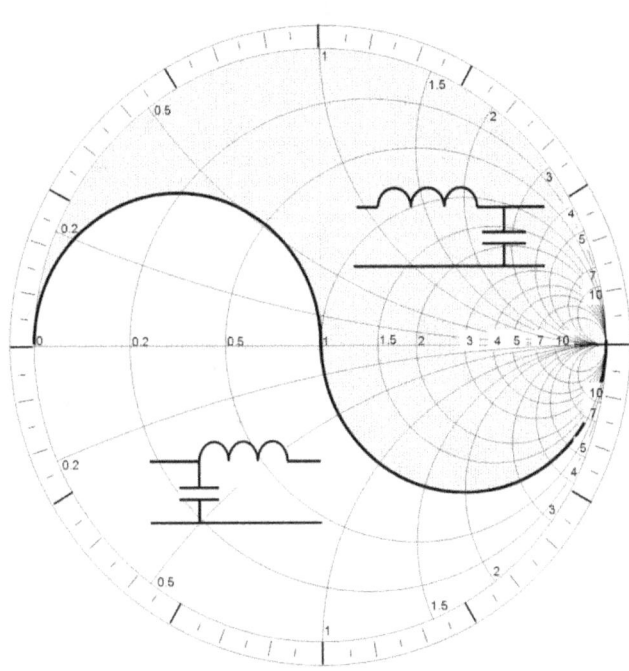

REVERSING THE L/C L NETWORK
FIGURE 6-9

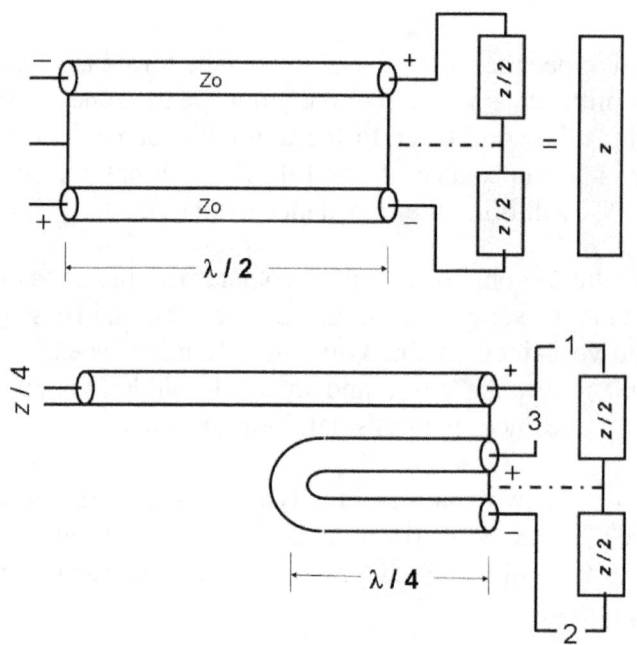

HALF-WAVE LINE 4:1 BALUN
FIGURE 7-1

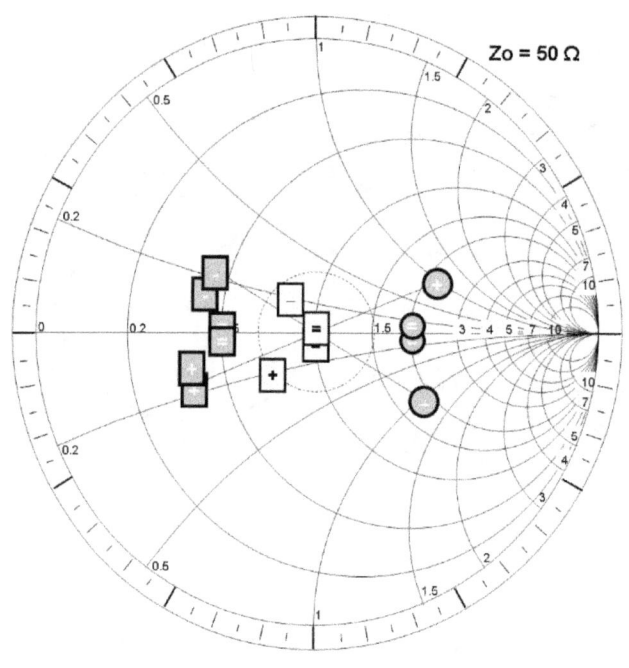

HALF WAVE LINE 4:1 BALUN
FIGURE 7-2

CABLE BALUNS

"Balun," as a term, has escaped definition. Technically, it is a contraction of the phrase *balanced to unbalanced*, used to define a two-port network that interconnects a balanced load, such as a dipole antenna, to an unbalanced coax cable. It does not imply an impedance transformation, but modern use of the term means any combination of impedance transformation and balanced to unbalanced interconnection.

The literature is awash with discussion of baluns. They probably started in the telephone and/or power industry a century or more earlier. I discuss only some concepts of using transmission lines to perform either or both balun functions.

Note that the Q section and sixty-degree baluns described in Chapter 5 are themselves cable baluns.

Half Wave Line

The first balun is a very common half-wave piece of coax. Half-wave dipoles generate a balanced load impedance of Z ohms. Let's normalize that to Z_o = 50 ohms. The load can be analyzed as two separate loads in series, each of impedance $z/2$, and each can be fed by a separate coaxial cable. The midpoint, the coax shields tied together, is at the same potential as the midpoint between the two loads, and in some antennas these are connected. Here they are left floating. This is the situation in the top of Figure 7-1.

The impedance at the left end of each coax can be found using a smith chart. It is a set of unbalanced impedances. At the left ends there's a 180° difference in phase of the voltage and current: the signs reverse as shown. The half wave coax impedance has a complete turn 'round the chart: the impedance $z/2$ repeats itself.

In Figure 7-1 he bottom coax has been folded back on itself into a "U." On this figure connect the upper and center coax ports together: they're in phase. This places two parallel loads, each of $z/2$ across the upper coax, for a combined load of $z/4$. The upper coax can be of any length and operate at a given VSWR which we can determine from the chart, once Z and Z_o are known. Interestingly, the lower coax can have any characteristic impedance! The balun has both converted a balanced load to an unbalanced cable, and achieved a 4:1 impedance reduction.

Here's how it works. Our subject folded dipole has Z = 200 ohms ($z = 4 + j\,0$) at frequency *f*, where it is resonant. Then, $z/2 = 2 + j\,0$. At 97% of *f*, $z/2 = 1.8 - j\,1$, and at 103%, $z/2 = 2.2 + j\,1$. The upper and lower limits are plotted on Figure 7-2 (previous page) as points "+" and "–" in shaded circles, respectively. Just either side of resonance, z/2 plots in similar circles labeled "=". These are the impedances at points *1* and *2* of Figure 7-1. What happens at point *3*?

It's in parallel with point *1* : we really want the admittance there, so convert to admittance mode. Rotate the four shaded circles half way 'round the chart to obtain the four shaded squares. At 103% of *f*, the U-shaped coax is too long: move the "+" square 5 degrees further clockwise to accommodate this, and by the same argument, the line is too short at 97% of *f* : move the "–" square 5 degrees counterclockwise. The "=" squares stay put. And at point *1* all squares stay put as well.

Add each pair of squares' admittances to obtain the four unshaded squares, the admittance seen at the end of the upper coax. Notice that all four points fall inside the dashed circle, marking VSWR = 1:5:1, and that the 200 ohm balanced load has been converted to a 50 ohm unbalanced load. Interesting …

Choke Baluns

A common balun is made by placing ferrite cores on the coax at the antenna. This forms an inductor, the outside shield of the coax being the inductor's wire.

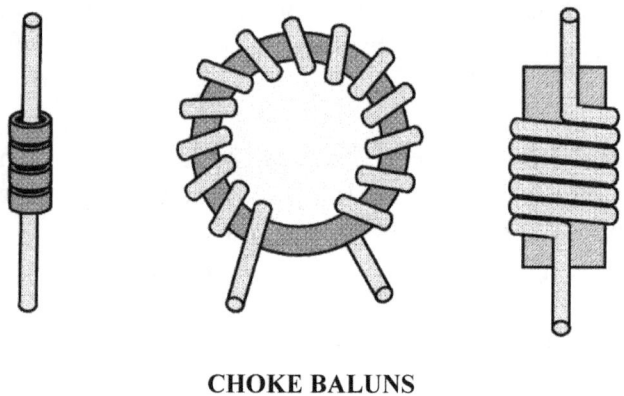

CHOKE BALUNS
FIGURE 7-3

On the left is an example of placing a stack of ferrite "beads" on coax: the center shows winding several turns of coax on a toroid core: the right shows an air core solenoid coil of coax, which may also contain ferrite rods to increase inductance. All put an inductance in series with the coax. Suppose the inductor is designed with a reactance of 600 ohms. The effect is to put a 600 ohm impedance in series with the coax cable at the antenna, which attenuates RF current flowing "on the outside" of the coax.

Does this stop the RF current traveling further down the coax? No – for there must be some current through the inductor to generate its voltage drop. Zero current would render the inductor invisible, RF-wise.

The solenoid core choke, however, *can* stop current on the coax. Because of capacitance between the windings, it becomes a parallel L/C circuit at some particular frequency and hence have nearly infinite impedance, aside from a bit of resistance in shunt across the coil.

But there's another way of generating a very high impedance choke on the coax.

Quarter-Wave Line

Use a piece of coax as if it were simply a thick wire of radius r, and bend it into a shorted, parallel wire, quarter wave stub resonant at frequency f. Zo of the stub is given as 276 times log(D/r) where D is the center-to-center spacing of the two pieces of coax, and r its diameter. The stub is shown in Figure 7-4 (next page). For cable such as RG-8, with radius of ¼ inch, a spacing D of ten times this, 2 ½ inches, Zo = 276 ohms. When calculating its length, its velocity factor is 100%, as the stub itself is air insulated.

At frequency f, it is a Q section, transforming the short circuit, shown as *S* on Figure 7-5, to an open circuit, infinite impedance, shown as "=". The stub is 90° long. But at $^2/_3 f$ it's 60° long with impedance Zo = 0 +j 467 ohms, and at twice that, at $^4/_3 f$, its impedance is 0 –j 467 ohms. Thus over one octave of frequency, it places at least 467 ohms in series on the coax run. This is a type of current choke, useful, say, from 14 to 28 MHz, or perhaps just a tad farther.

Two such chokes can be placed in series on a coax, one twice as long as the other, to extend the frequency range to two octaves. A third can extend it to three octaves, such as 3.5 to 28 MHz   Note that these are 1:1 baluns and do not change the load impedance.

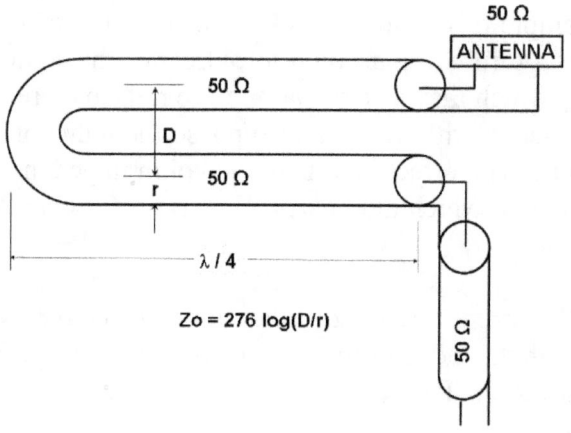

QUARTER WAVE LINE BALUN
FIGURE 7-4

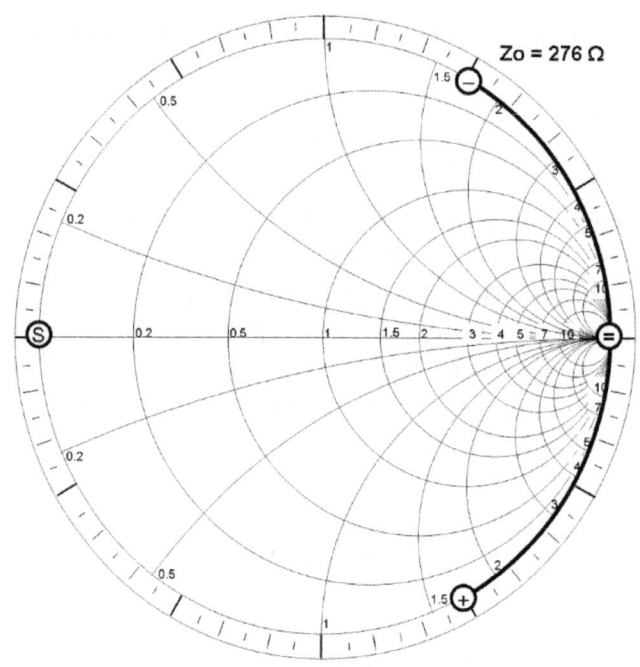

ANALYSIS
FIGURE 7-5

NOTES

Thin Dipole				Zo 50 ohms				
				DIPOLE IMPEDANCE			REF COEFF	
kHz	length degrees	R ohms	X ohms	z = r+ jx r	x	|z|	Re(p)	Im(p)
1900	45,6	5	-1600	0,10	-32,00	32,00	1,00	-0,06
3700	88,8	10	-440	0,20	-8,80	8,80	0,97	-0,22
5300	127,2	24	-240	0,48	-4,80	4,82	0,88	-0,38
7125	85,5	70	0	1,40	0,00	1,40	0,17	0,00
10100	171,0	100	380	2,00	7,60	7,86	0,91	0,23
14200	341,0	1500	-600	30,00	-12,00	32,31	0,94	-0,02
18100	434,0	120	-300	2,40	-6,00	6,46	0,86	-0,25
21200	508,0	80	-100	1,60	-2,00	2,56	0,52	-0,37
24900	599,0	150	200	3,00	4,00	5,00	0,75	0,25
29000	696,0	1500	0	30,00	0,00	30,00	0,94	0,00

TABLE 8-1

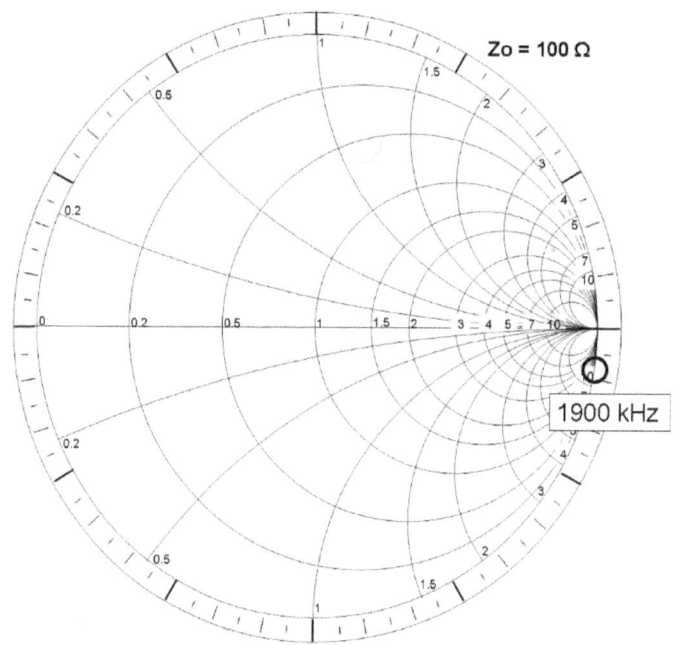

- 8 -

ZEPPS AND J POLES

Center Fed Antenna

Begin with a thin wire, dipole antenna, a straight wire cut in the middle where transmission line is attached. We'll study one designed for the amateur radio 40 meter band, 20 meters long, resonant at 7125 kHz.[8] This is a classic centerfed zepp. It can be operated at considerably lower or higher frequencies, such as 1.8 to 30 MHz. From Brown and Woodward, the dipole's impedance vice frequency is given in Table 8-1 for lengths less than 21 MHz. I've added reasonable "guesstimates" for higher frequencies: the report does not extend upward that far.

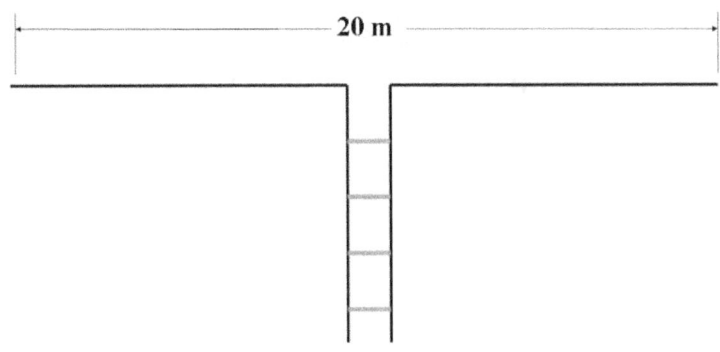

HALF WAVE CENTER FED ZEPP, 7125 kHz
FIGURE 8-2

Operation Near ¼ Frequency

At 1900 kHz the normalized impedance has magnitude $|z| = 32$ with $Z_o = 50$ ohms, so $|Z| = 1600$ ohms. If I could feed it current at 100 mA, the power input would be 16 watts, highly reactive. Its low resistance of 5 ohms dissipates only 50 mW (current squared times resistance) as power loss from radiation plus antenna heating. The rest reflects back down the transmission line.

A shorted quarter wave line, driven with some power, develops a high voltage at its open end and a high current at the other. The open end can drive voltage into the dipole. However, the maximum current will be at the feedpoint, and there will be less along its length. The antenna will not be a good radiator.

---

[8] ½ x 300/f(MHz) x 95%

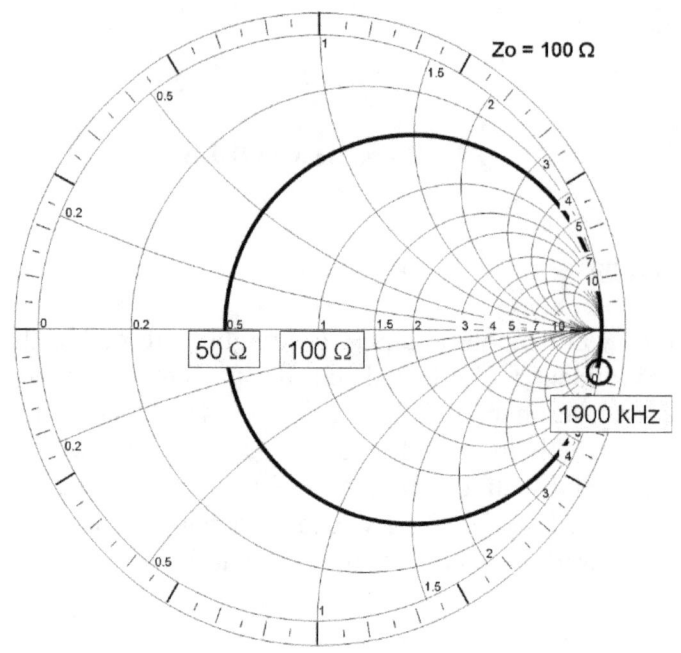

50 OHM CIRCLE ON 100 OHM LINE
FIGURE 8-3

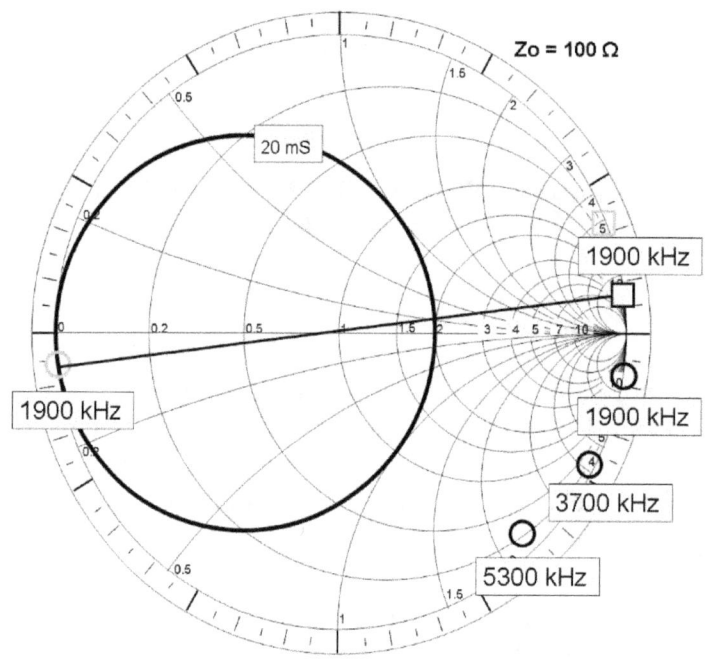

20 mS CIRCLE
FIGURE 8-4

We could tune this antenna by the series – parallel line technique of Chapter 6, using parallel, open-wire feedline that operates at a very high VSWR. Customarily one uses very low-loss line for this, and historically open-wire line of 400 to 600 ohms has been used. But my argument is better demonstrated with lower impedance line, otherwise the impedance and admittance points on the smith chart fall too close to the outer circle and it is difficult to discern the actual action. I'll use 100 ohm line and match the antenna to a 50 ohm source.

Figure 8-1 (page 42) plots the 1900 kHz impedance, showing a difficult situation. The impedance is way out toward the chart edge, near $z = -j\,16$. The objective is to bring this to a standard value, 50 ohms, non-reactive. On Figure 8-3 the 50 ohm circle appears, and on Figure 8-4, its companion 20 mS circle, along with impedances for the three lowest frequencies of the table.

Following Chapter 6 with a slight change of target susceptance circle, a series line can bring the 1900 kHz point to the target 20 mS circle, near $x = -0.05$. The length needed is about 75°, or 33 meters. At this point a shunt piece of line can be attached, shorted at its end, about 8 degrees long, as seen by the companion admittance square on the right hand side of Figure 8-4. The impedance at the series/shunt juncture will be 50 ohms.

Action At Somewhat Higher Frequencies

The antenna can be tuned at 3700 kHz with a series line 60° long – about 13.5 meters – and a short shunt stub, and again at 5300 kHz with a series line 68° long – about 10.7 meters – and companion shunt stub. And at 7125 kHz, the antenna is resonant with an impedance of $70 +j\,0$ ohms. It needs no additional tuning.

There are frequencies whose impedance falls inside the VSWR 2:1 circle. These are near $y = 1 + j0$. No amount of distance down the line moves them to the target susceptance circle. The situation is shown in Figure 8-5 (next page). For these, plot their admittance instead of impedance: the solid square.

Move along the line any distance. The square simply runs around at constant VSWR, inside the 2:1 circle. Here, the square rests at $y = 0.8 +j\,0.3$. Move it at constant conductance to the 2:1 circle at $y = 0.8 -j\,0.6$ by a shorted shunt stub, with $b = -0.9$.

Now continue along the line about 138 degrees clockwise, and the admittance will become $2 +j\,0$. The line Zo is 100 ohms: the chart center is 10 mS, and the admittance is now 20 mS, with an impedance of 50 ohms as desired.

This is a bit of a variance from Chapter 6, brought about by use of a line with higher than 50 ohm impedance, to match to 50 ohms.

While in principle a center fed zepp antenna can be tuned with series and shunt transmission lines, this is usually not done. Instead a line of convenient length is brought to the radio shack where an agile antenna tuner finishes impedance matching to the radio gear. This works well with balanced antennas.

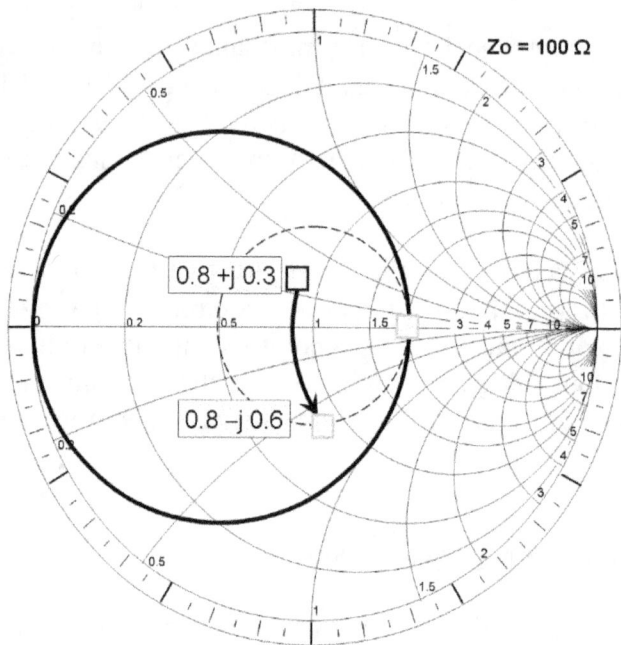

SHUNT / SERIES STUB TUNING
FIGURE 8-5

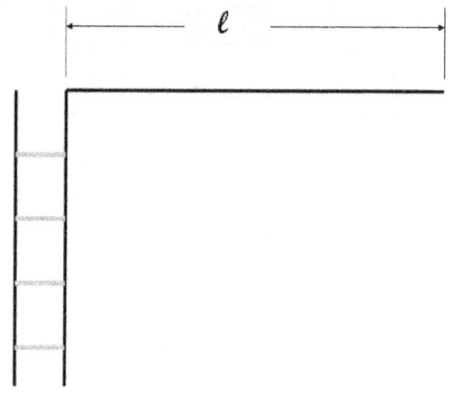

END FED ZEPP ANTENNA
FIGURE 8-6

End-Fed Zepp

But another variant is the unbalanced end-fed zepp antenna, where one of the two dipole arms is missing, as in Figure 8-6.

One end of the transmission line is left open. No current can flow out of it. The other end provides current, giving rise to the unbalance condition. The line then operates in two modes: its usual, balanced or *differential* mode with equal but opposite currents flowing in the two halves, and an unbalanced, *common* mode, where currents in the two halves are in, not out, of phase.

The differential mode doesn't radiate, except for a tiny bit in the plane of its two wires, while the common mode radiates as if it were a thick piece of ordinary conductor.

End fed zepps require minimum common mode current, achieved by choosing its length, ℓ, to present the highest possible impedance. This happens a tad shy of multiples on one half wavelength, the half wave case being a common value.

The J-Pole

This rather popular antenna is but a variant of the classic halfwave end fed zepp. The half wave section is simply an extension of half the feedline, and the antenna is most often oriented along a vertical axis. It radiates broadside to the half wave.

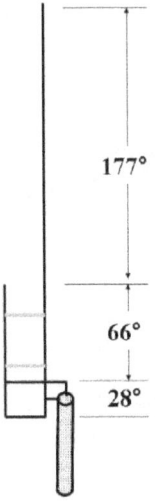

**J – POLE ANTENNA
FIGURE 8-7**

In a typical design, a half wave vertical, 177° tall, presents an impedance of 1000 +j 0 ohms,. On 100 ohm balanced line, $z = 10 + j\ 0$ as plotted on Figure 8-8, right side of the real axis.

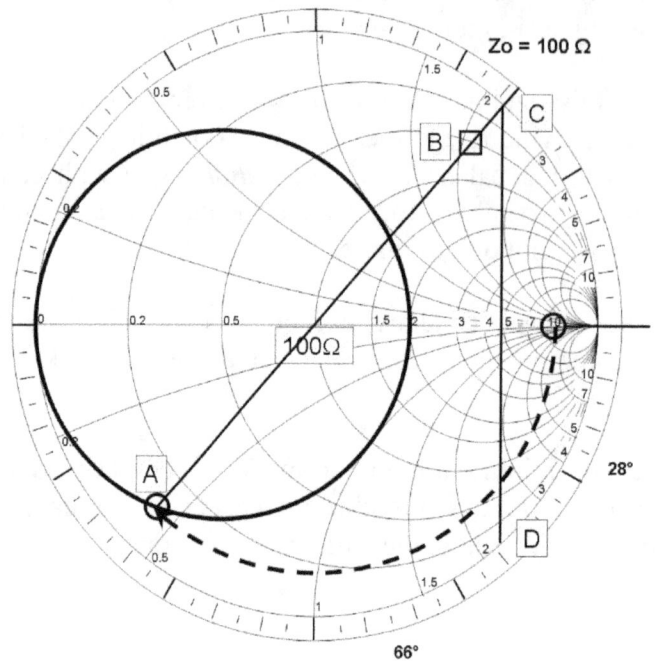

J-POLE DESIGN
FIGURE 8-8

Proceed clockwise 66° at constant VSWR to the image of the 20 mS circle, at point **A**. Then convert to admittance by reflecting through the chart center to point **B**: $y = 0.5 +j\ 2.2$. A shorted 28° stub will provide $y = 0 -j\ 2.2$., leaving the required impedance of 0.5 +j 0, 20 mS, or 50 ohms impedance.

The overall antenna then has a height of 177° + 66° + 28°, or 271° – almost exactly ¾ wavelength. Hence the antenna is often thought of as a half wave radiator above a quarter wave shorted stub, as seen on the figure.

Two Problem Areas

First: radiation from the quarter wave matching line modifies the antenna pattern: it lifts the main lobe about ten degrees above the horizon, with maximum gain along the horizon reduced a small amount – less than a dB. This effect can be mitigated by laying the matching section horizontally, or wrapping it into a small horizontal circle.

Second: The coax line can itself carry unbalanced current and radiate, and this effect varies with the length of the coax. Avoid this with a choke balun of some sort, placed on the coax not more than 1/8 wavelength from the antenna. This induced high impedance point on the coax will prevent it from becoming near resonance and carrying appreciable common mode current.

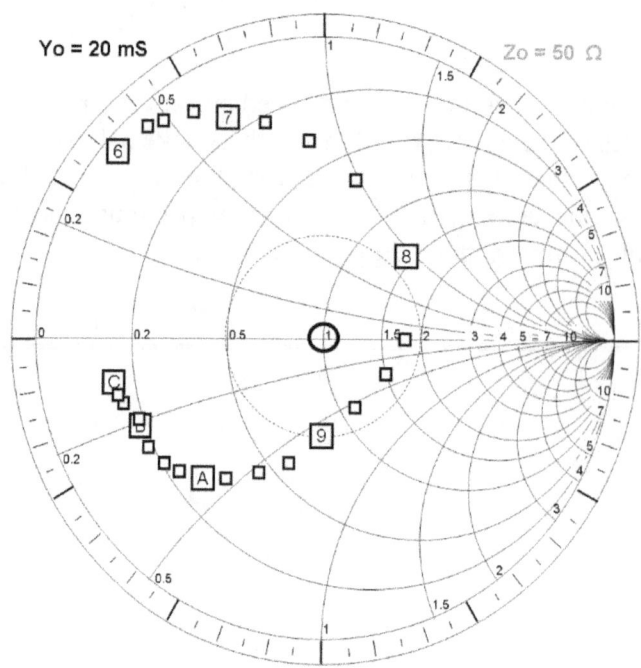

INPUT ADMITTANCE – MONOPOLE ANTENNA
FIGURE 9-1

chart notation	height degrees	z = r+ jx r	x	y = g+ jb g	b	Re	Im
6	60,0	0,16	-2,70	0,02	0,37	0,73	-0,63
	62,5	0,18	-2,20	0,04	0,45	0,62	-0,71
	65,0	0,22	-2,00	0,05	0,49	0,56	-0,73
	67,5	0,26	-1,72	0,09	0,57	0,45	-0,76
7	70,0	0,34	-1,50	0,14	0,63	0,34	-0,74
	72,5	0,38	-1,26	0,22	0,73	0,21	-0,72
	75,0	0,42	-0,98	0,37	0,86	0,05	-0,66
	77,5	0,46	-0,70	0,66	1,00	-0,11	-0,53
8	80,0	0,50	-0,32	1,42	0,91	-0,28	-0,27
	82,5	0,56	0,00	1,79	0,00	-0,28	0,00
	85,0	0,64	0,16	1,47	-0,37	-0,21	0,12
	87,5	0,72	0,36	1,11	-0,56	-0,11	0,23
9	90,0	0,80	0,60	0,80	-0,60	0,00	0,33
	92,5	0,84	0,88	0,57	-0,59	0,12	0,42
	95,0	0,92	1,10	0,45	-0,53	0,22	0,45
	97,5	1,00	1,40	0,34	-0,47	0,33	0,47
A	100,0	1,10	1,70	0,27	-0,41	0,42	0,47
	102,5	1,24	2,00	0,22	-0,36	0,50	0,44
	105,0	1,40	2,24	0,20	-0,32	0,55	0,42
	107,5	1,56	2,56	0,17	-0,28	0,61	0,39
B	110,0	1,70	2,80	0,16	-0,26	0,64	0,37
	112,5	1,86	3,04	0,15	-0,24	0,67	0,35
	115,0	2,10	3,28	0,14	-0,22	0,70	0,32
	117,5	2,40	3,60	0,13	-0,19	0,72	0,29
C	120,0	3,00	3,80	0,13	-0,16	0,74	0,25

MONOPOLE, NORMALIZED TO Z_0=50 OHMS
TABLE 9-1

MONOPOLE AND DIPOLE SHUNTS

Ordinary Monopole

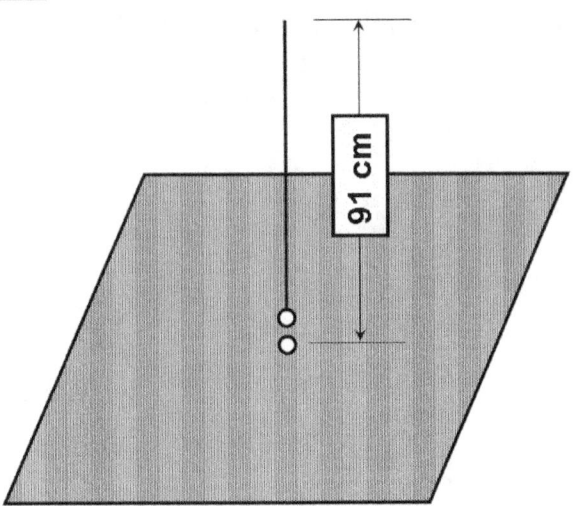

90 DEGREE, 90 MHz MONOPOLE
FIGURE 9-2

Brown and Woodward's seminal report[9] presents data on input impedance of monopole antennas above a large ground plane. On Figure 9-1 I've plotted these, spun half way 'round the chart to show input admittance. Notations **6** through **C** refer to monopole heights of 60°, 70°, 80°, 90°, 100°, 110°, and 120° respectively. The 2:1 VSWR circle is also shown. The normalized admittance y has conductance g, susceptance b: $y = g + j\,b$. I'll use those notations below.

Study a monopole for 90 MHz: the angle notations then refer to the frequency driving the antenna. The figure shows that, for 82 to 90 MHz, y falls in the 2:1 circle: this is the antenna's frequency range, a bandwidth of 8 MHz. Its Quality Factor, Q, is defined as one-half the sum over the difference, and its bandwidth, the inverse of Q times 100%. These are specified at VSWR not over 2:1.

$$Q = \tfrac{1}{2}\,(82+90)\,/\,8 = 10.75. \qquad 1\,/\,Q \times 100\% = 9.3\%.$$

How can its bandwidth be increased?

---

[9] "Experimentally Determined Impedance Characteristics of Cylindrical Antennas," George H. Brown and Woodward, O. M., Proc. of the IRE, April 1945.

Below 82.5 MHz the antenna is capacitive, and above, inductive. A shunt with b going the other way could do the trick. From Figure 9-1 there's hope for 77.5 to 92.5 MHz, as these are inside the $r = 0.5$ circle. Shunt in b might move their points along lines of constant g toward the 2:1 circle.

A shorted, 50 ohm, 90 MHz. Q-section has b = zero at point **9** on Figure 9-3. Elsewhere it has proportionally more or less, as in Table 9-2.

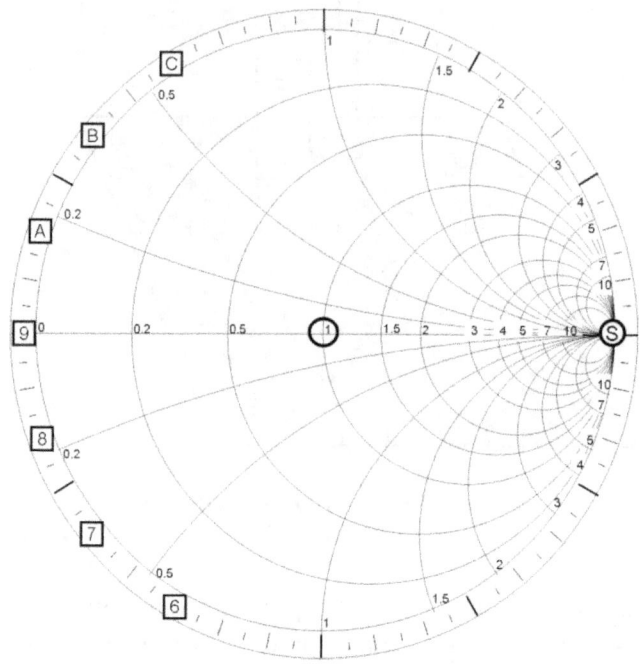

QUARTER WAVE SHORTED STUB - SUSCEPTANCE vs LENGTH
FIGURE 9-3

70		75		80		85		90		95		100
-0,36	-0,32	-0,27	-0,22	-0,18	-0,13	-0,09	-0,04	0,00	0,04	0,09	0,13	0,18

50-OHM SHORTED STUB NORMALIZED SUSCEPTANCES
TABLE 9-2

Let's combine Tables 9-1 and 9-2 for these two frequencies.

9.1	75 MHz	+0.37 + j 0.86	95 MHz	+0.45 − j 0.53
9.2		−j 0.27		+j 0.09
SUM		+0.37 +j 0.59		+0.45 − j 0.44

52

These are improvements. The antenna now has less susceptance. All other admittances are likewise changed. Figures 9-4 (unmodified antenna) and 9-5 (next page – stub shorting the antenna) show this improvement. In the latter, the frequency range has increased to 80 to 90 MHz:

$Q = ½ (80+90) / 10 = 8.5$ $1 / Q \times 100\% = 11.8\%$.

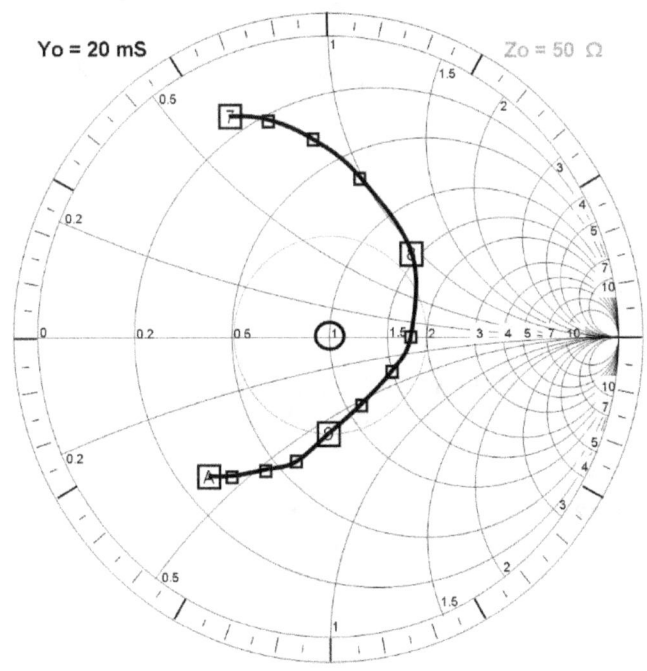

ANTENNA ADMITTANCE, 70-100 MHz
FIGURE 9-4

Well: if one stub is good, two might be better. Together two just double the *b* of the stub, as in the calculation below. Figure 9-6 (next page) shows the effect.

9.1	75 MHz	+0.37 +j 0.86	95 MHz	+0.45 −j 0.53
9.2		−j 0.54		+j 0.18
SUM		+0.37 +j 0.32		+0.45 +j 0.36

These change *b* at 75 and 95 MHz to +0.32 and +0.36 respectively. The 2:1 frequency range is seen to be 78 to 90 MHz:

$Q = ½ (78+90) / 12 = 7.0$ $1 / Q \times 100\% = 14.3\%$.

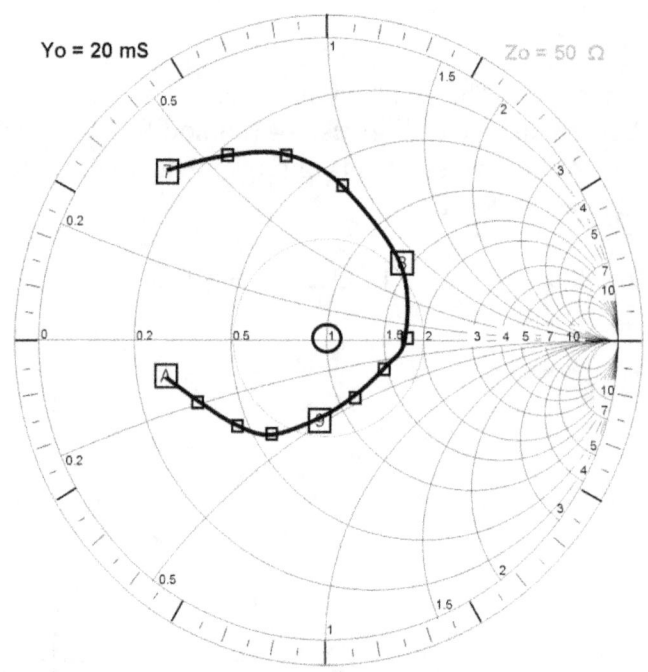

ADMITTANCE WITH 50 OHM STUB
FIGURE 9-5

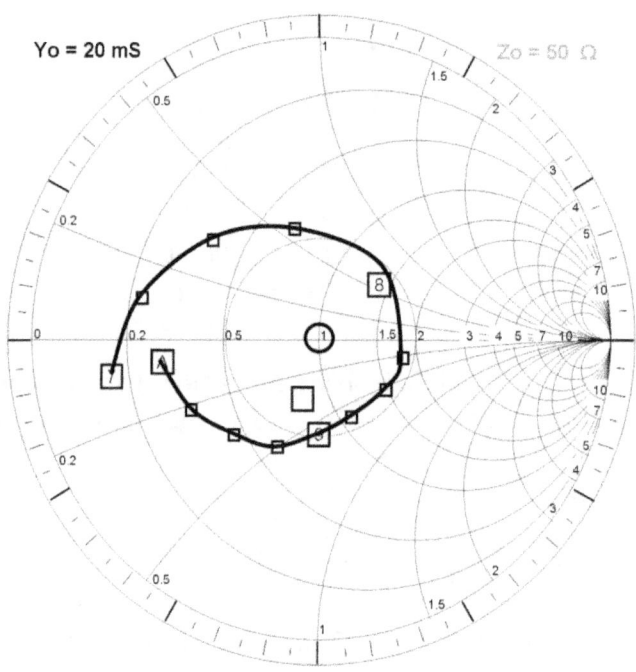

ANTENNA WITH TWO 50 OHM STUBS
FIGURE 9-6

Adding a third stub produces the effect shown below in Figure 9-7, with a frequency range of 77 to 90 MHz:

$Q = ½ (77+90) / 12 = 6.42$ $1 / Q \times 100\% = 15.57\%$.

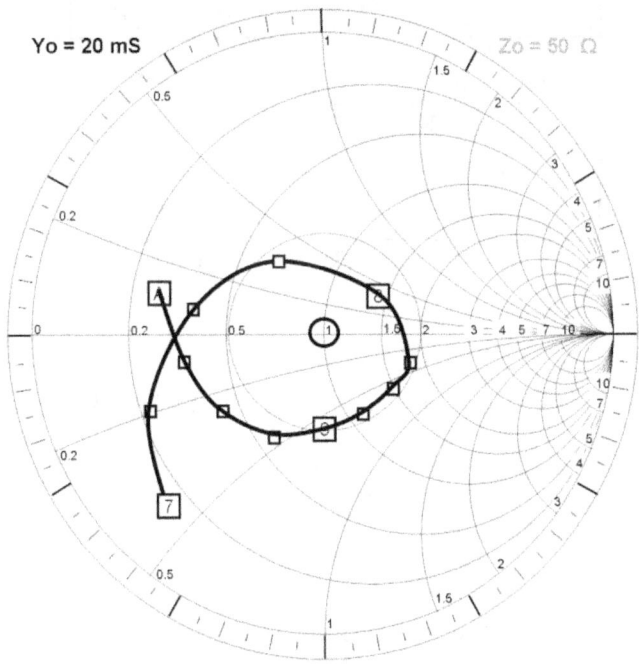

ANTENNA WITH THREE FIFTY OHM STUBS
FIGURE 9-7

A drawing of the triply-shunted monopole so designed appears below, Figure 9-8.

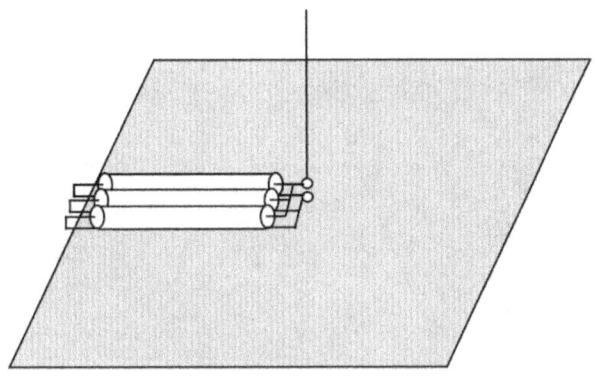

SHUNTED MONOPOLE ANTENNA
FIGURE 9-8

The VSWR without and with three stubs is shown in Figure 9-9.

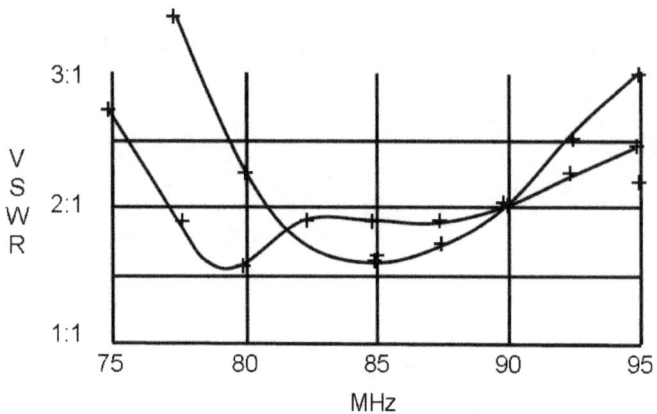

**VSWR CURVES
BEFORE (UPPER) AND AFTER (LOWER)
FIGURE 9-9**

Folded Monopole

The three shunt stubs of Figure 9-8 operate as classic coax, in *differential* mode. Current in the center conductor is exactly balanced by countercurrent "inside" the shield. But in many designs the vertical element is folded back on itself, as shown in Figure 9-10, left.

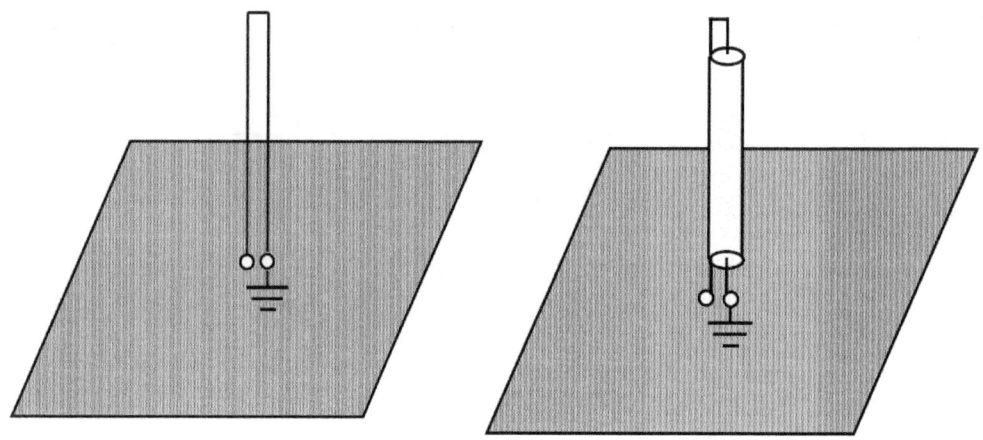

**FOLDED MONOPOLE DESIGNS
FIGURE 9-10**

There are also designs where the vertical element is a quarter wave Q-section, as on the right. A common case of the latter is a class of AM broadcast towers, the tower itself being the grounded, center conductor and a cage of six or more vertical wires dropped from crossarms at the top being the outer conductor of unbalanced, coax-like cable.

In such designs, the Q-section operates in two modes: *differential* mode effecting the shunting stub, and *common* mode (currents on inner and outer conductors now in phase – i.e., current on the "outside") doing the actual radiating.

Although the same principles of shunt stub design apply, there's a change in impedance of the antenna. The two-equal wire version on the left has impedance four times that of a single monopole. Zo of the stub must be changed accordingly.

Folded Dipole

A dipole is just two monopoles, back to back. Figure 9-11, left, depicts this, with a large metal sheet between the two. But the sheet can be removed and the same electric and magnetic fields remain: the common arrangement on the right results. In these designs, the folded dipole input impedance is four times that of a single dipole, so the stub impedances need to be changed. It's interesting that the dipole's two arms are themselves Q-section stubs and, if designed properly, provide the broadbanding effect discussed in this chapter. Hence: a folded dipole's 2:1 bandwidth is greater than that of straight dipoles.

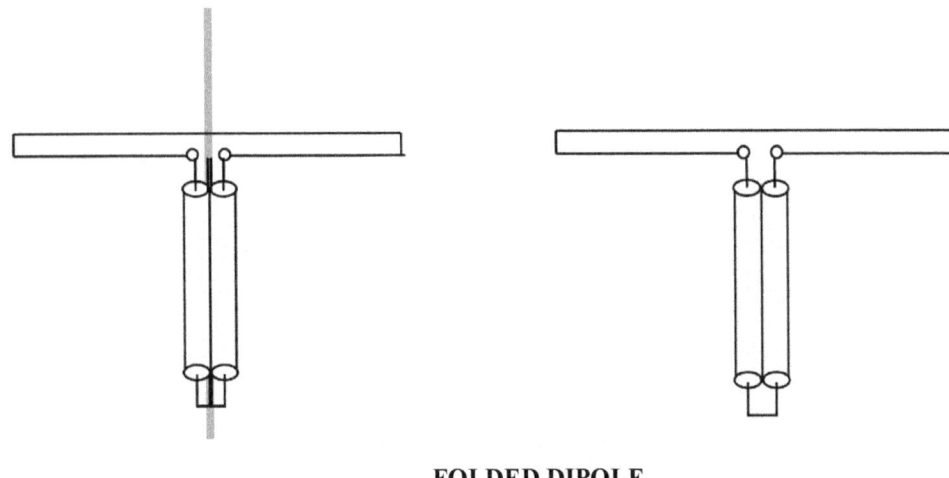

FOLDED DIPOLE
FIGURE 9-11

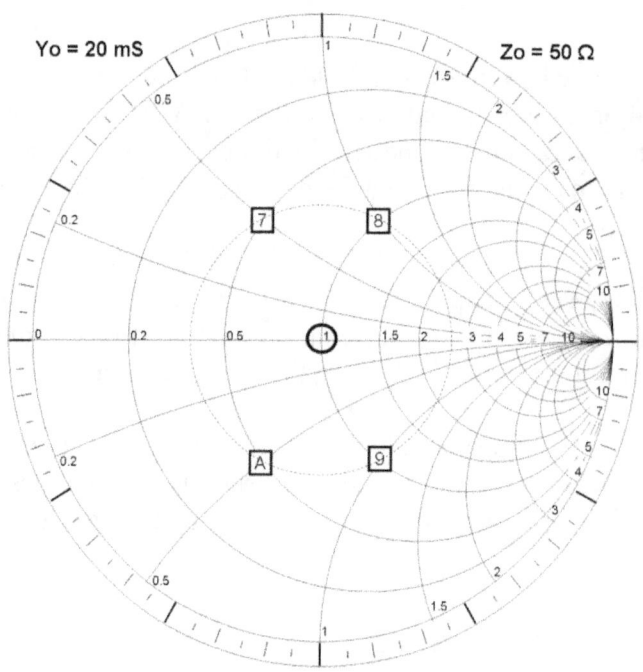

A SYNTHESIZED ADMITTANCE PLOT
FIGURE 10-1

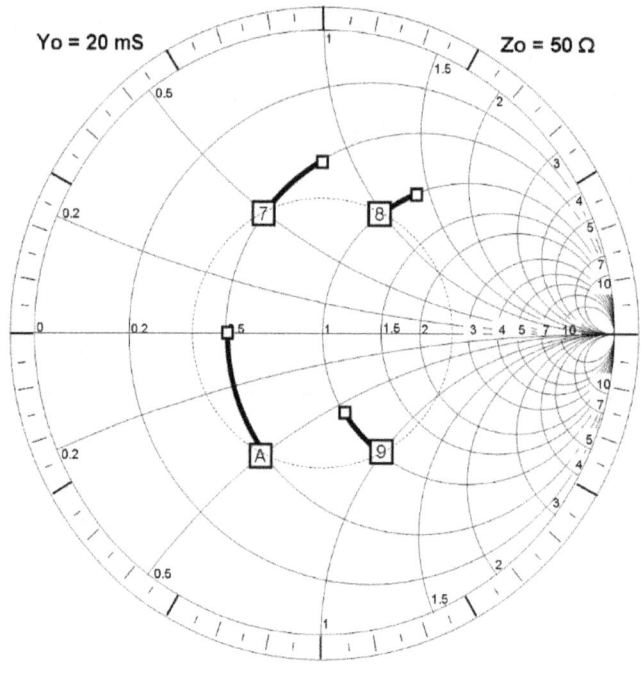

EFFECT OF SHUNT CAPACITANCE
FIGURE 10-2

CONCEPTS OF BROADBANDING

Many antennas have a high Q and limited bandwidth. Broadbanding is the art of increasing the latter. By example, radio amateurs are permitted to use the 1800 to 2000 kHz band. An antenna for same offering a 2:1 maximum VSWR will have a Q not exceeding ½(1800 + 2000) / 200 = 9.5, a bandwidth of 10.5%. The average dipole antenna has a bandwidth of about 9.3%, a Q of 10.75, according to Chapter 9. It cannot support operation on that entire band without retuning.

Chapter 9 has shown how common stub tuning can reduce Q to below 6.5. This chapter explores the concept of increased bandwidth in more detail by further study of that chapter's monopole.

As was previously noted, the impedance of an antenna changes with distance along the coax toward the radio gear. We can cut the coax at judiciously chosen spots and insert series impedance there. Done properly, the VSWR can be lowered. However, I **do not** further develop that theme. Instead, this study explores insertion of shunt **admittances** at those spots, where the shunt is restricted to non-lossy, pure susceptances.[10]

Shunt Capacitance

Begin with a synthesized admittance plot, Figure 10-1. Four points are shown, again for 70, 80, 90 and 100 MHz, as points *7, 8, 9,* and *A*, respectively. When shunt capacitance is added, the admittances move along lines of constant conductance, seen in Figure 10-2. Capacitive susceptance is proportional to frequency:[11] as shown, the shift at 100 MHz is to add susceptance $b = 0.5$, 0.45 at 90 MHz, 0.4 at 80 MHz, and 0.35 at 70 MHz.

The VSWR 2.6:1 circle appears on the plot. Adding capacitance is seen to increase VSWR below 85 MHz, and to reduce it above. Clearly more compensation is needed. But this stratagem can move the entire plot in the upward direction and twist it a bit clockwise.

---

[10] I.e, capacitive or inductive shunts without resistance.
[11] $X_C = 1/2\pi fC$ $B_C = 1/X_C = 2\pi fC$ $X_L = 2\pi fL$ $B_L = 1/2\pi fL$

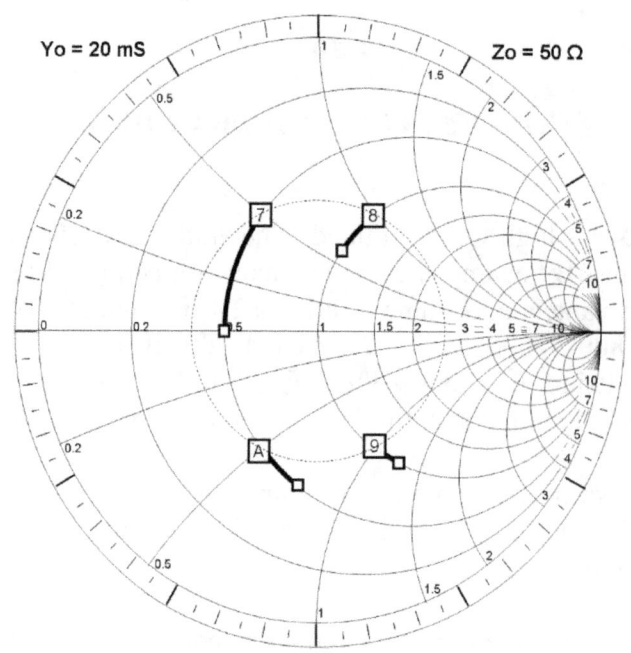

EFFECT OF SHUNT IDUCTANCE
FIGURE 10-3

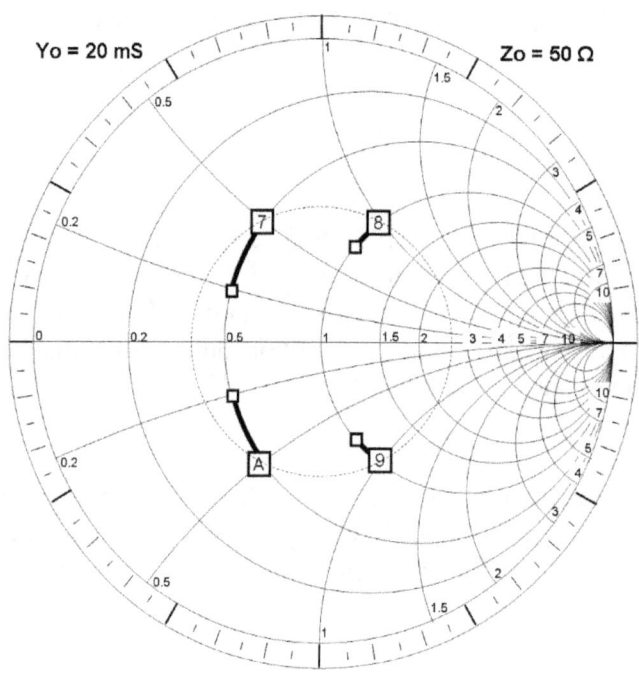

EFFECT OF SHUNT STUB
FIGURE 10-4

Shunt Inductance

Shunt inductance moves the susceptances again along lines of constant conductance, but in the other direction, as in Figure 10-3. Now the change in b is inversely proportional to frequency: at 70 MHz $b = -0.5$: at 80 MHz $= -0.44$, at 90 MHz $= -0.39$, and 100 MHz $= -0.35$ as shown. Now the VSWR below 85 MHz is reduced, and above, increased. Again more or differing compensation is needed. This stratagem, however, can move the entire plot downward on the chart and twist it a bit counter-clockwise.

Shunt Stub

Stub tuning was previously encountered in Chapter 9. Make a stub, shorted at its far end, a quarter wave long at 85 MHz: its susceptance there is zero. Its Zo is chosen to provide inductive $b = -0.3$ at 70 MHz, and capacitive $b = +0.3$ at 100 MHz. This moves all points closer to the real axis if not overdone, and reduces VSWR everywhere. See Figure 10-4. We seem to be on the right path.

Change of Zo

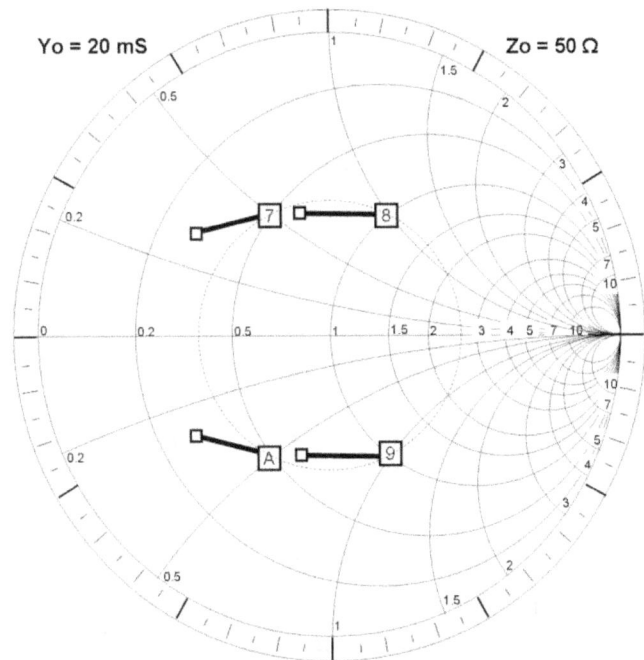

EFFECT OF CHANGING LINE Zo
FIGURE 10-5

The impedance of transmission line can be changed. Figure 10-5 (previous page) shows a change from an initial value of 50 ohms (Yo = 20 mS) to 30 ohms (Yo = 33 mS). This change shifts the entire pattern to the left. An increase in Zo shifts it to the right.

Rotation

A move along the transmission line moves all admittance points clockwise at constant VSWR. The move is proportional to frequency. Figure 10-6 shows the effect of a rotation of 20° electrical length at 100 MHz. The rotation at 90 MHz is 18°, 16° at 80 MHz, and 14° at 70 MHz.

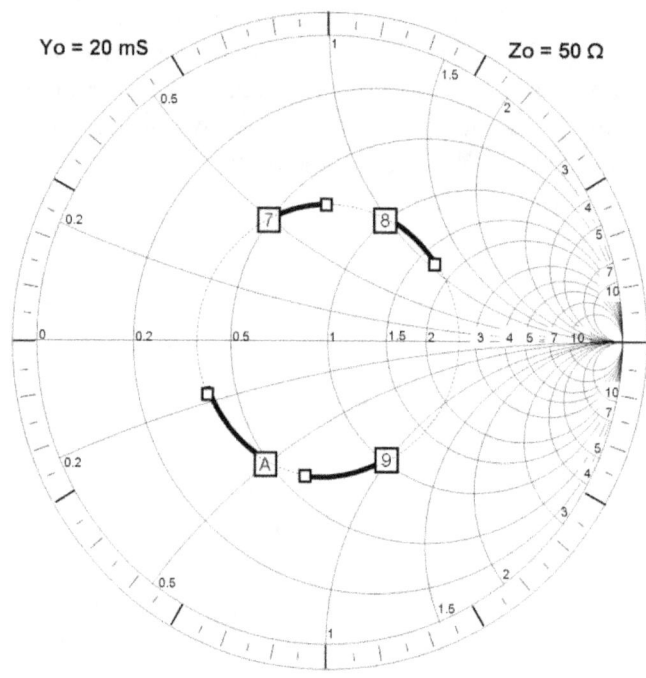

**ROTATION
EFFECT OF MOVING ALONG THE LINE
FIGURE 10-6**

These five tools: capacitive shunt, inductive shunt, stub shunt, change of Zo, and rotation by a move along the line toward the radio gear, grant a good deal of flexibility in reducing the VSWR of an antenna, lowering its Q, and increasing its bandwidth. We'll study them in more detail, using the monopole data introduced in Chapter 9.

The Study Antenna

Figure 10-7 plots the antenna's admittance, a repeat of Figure 9-4 in a slightly different form. It includes a rectangular scale of the reflection coefficient. Notations are again for 70, 80, 90 and 100 MHz: *7, 8, 9,* and *A*, respectively. The maximum VSWR, at *7*, is 9.8:1.

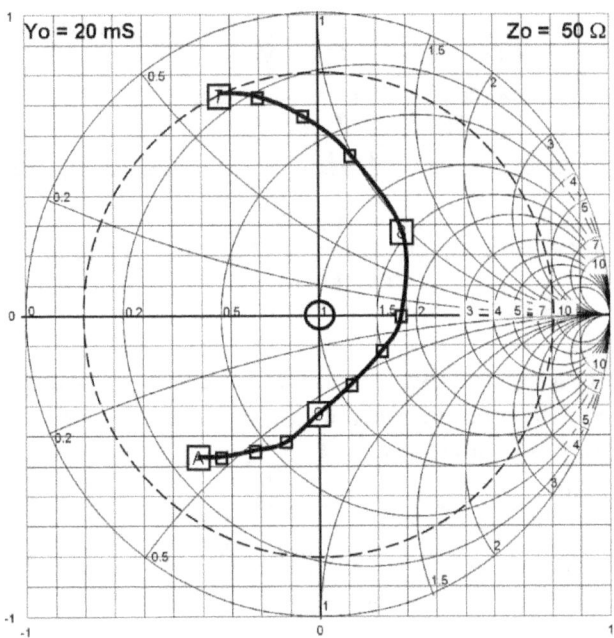

STUDY MONOPOLE ADMITTANCE
FIGURE 10-7

Change of Zo

I can change the impedance of the line. Suppose I change it from 50 to 30 ohms. The center point susceptance then changes from 1/50 = 20 mS to 1/30 = 33 mS. All points then shift to the left as described in association with Figure 10-5 above.

The effect could well be calculated by very careful scaling on a smith chart. However, I've done all the necessary calculations on a spreadsheet. Table 10-1 tabulates the change in reflection coefficient as I proceed in a step by step analysis, working toward an acceptable value of VSWR. The spreadsheet does the math, generating a blizzard of numbers whose results are rather difficult to track and "see." The smith chart, on the other hand, visually captures each change step by step and allows the designer to watch the process in pictorial form.

FREQUENCY	MONOPOLE REFL COEFF		CHANGE Zo TO 30 OHMS		STUB 30 OHM 89°		ROTATE 62°		SHUNT L b = -0.865 @ 90°		ROTATE 60°		SHUNT C b = +0.6 @ 90°		STUB 17 OHM 80°	
70,0	-0,34	0,74	-0,64	0,57	-0,84	0,00	0,10	0,84	-0,68	0,01	0,04	0,68	0,31	0,72	-0,28	0,46
72,5	-0,21	0,72	-0,54	0,59	-0,75	0,16	0,29	0,71	-0,34	0,33	0,37	0,30	0,46	0,36	0,29	0,24
75,0	-0,05	0,66	-0,39	0,59	-0,58	0,30	0,43	0,49	0,11	0,31	0,29	-0,16	0,25	-0,03	0,32	-0,22
77,5	0,11	0,53	-0,21	0,52	-0,35	0,35	0,43	0,23	0,34	0,05	-0,03	-0,35	-0,14	-0,05	0,03	-0,42
80,0	0,28	0,27	0,01	0,29	-0,04	0,20	0,20	-0,03	0,32	-0,29	-0,37	-0,22	-0,34	0,28	-0,34	-0,29
82,5	0,28	0,00	0,03	0,00	0,04	-0,07	-0,08	-0,01	0,14	-0,44	-0,46	0,01	-0,25	0,51	-0,46	0,03
85,0	0,21	-0,12	-0,05	-0,12	-0,03	-0,18	-0,14	0,11	0,02	-0,42	-0,39	0,14	-0,12	0,56	-0,37	0,25
87,5	0,11	-0,23	-0,15	-0,23	-0,14	-0,27	-0,16	0,25	-0,11	-0,35	-0,27	0,25	0,02	0,57	-0,19	0,39
90,0	0,00	-0,33	-0,28	-0,31	-0,27	-0,32	-0,12	0,40	-0,22	-0,22	-0,08	0,30	0,17	0,53	0,03	0,44
92,5	-0,12	-0,42	-0,39	-0,37	-0,41	-0,35	-0,03	0,53	-0,30	-0,03	0,14	0,27	0,31	0,42	0,24	0,38
95,0	-0,22	-0,45	-0,48	-0,37	-0,52	-0,30	0,11	0,59	-0,25	0,18	0,29	0,10	0,37	0,23	0,34	0,20
97,5	-0,33	-0,47	-0,58	-0,37	-0,63	-0,23	0,27	0,61	-0,11	0,39	0,37	-0,17	0,32	-0,03	0,33	-0,05
100,0	-0,42	-0,47	-0,65	-0,35	-0,71	-0,14	0,44	0,58	0,13	0,50	0,27	-0,44	0,08	-0,26	0,09	-0,27

**SPREADSHEET RESULTS
REFLECTION COEFFICIENT AT EACH STEP
TABLE 10-1**

Working on this problem quickly reveals that the bulge toward the right at the midrange of frequencies becomes troublesome. This bulge is removed by the stratagem of changing the feedline impedance from 50 to 30 ohms as in Figure 10-8. All points shift to the left.

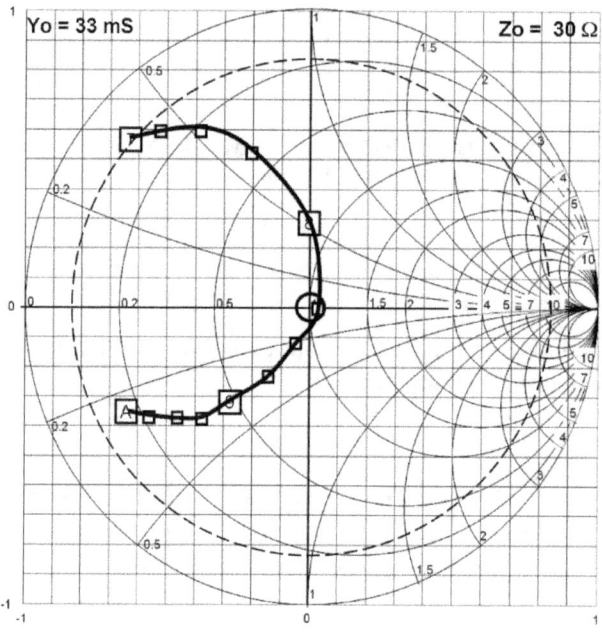

**CHANGE LINE Zo TO 30 OHMS
FIGURE 10-8**

But alas! The maximum VSWR has *increased* to 13.3:1! To reduce VSWR, I need to move point *7* to a higher susceptance circle. The next step, then, is to add a shunt stub. One made of cable with Zo = 30 ohms, 89° long at 90 MHz., changes the plot to that shown in Figure 10-9.

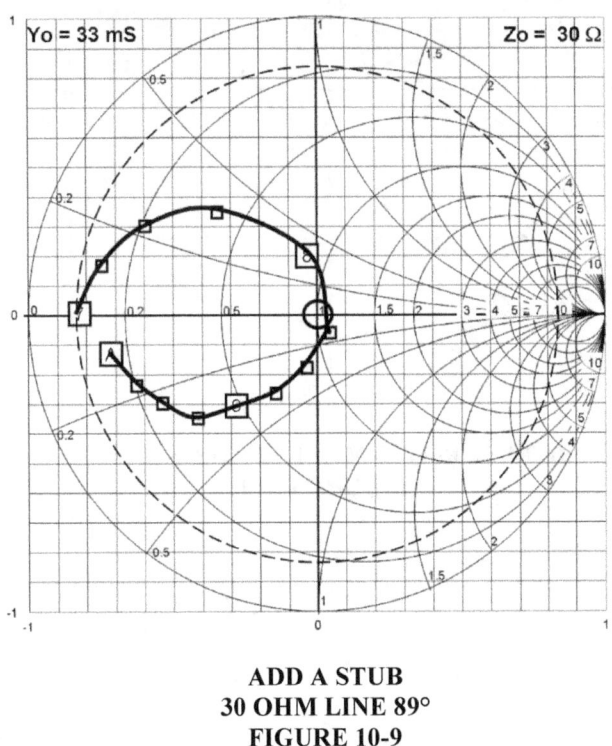

ADD A STUB
30 OHM LINE 89°
FIGURE 10-9

The VSWR has dropped a bit, to 11.6:1. I'm still worse off than when I started. I need to get point *7* to a still higher susceptance circle.

The next step, then, is to rotate the plot by moving further along the transmission line toward the radio gear. A 62° rotation produces Figure 10-10 (next page) But rotation moves all points along lines of constant VSWR – the maximum VSWR remains at 11.6:1. Obviously more compensation is needed.

Figure 10-10 begs that the plot be lowered on the chart. The way to do this is by adding a shunt inductor. Figure 10-11 (next page) shows the result of adding an inductive shunt, susceptance $b = -0.865$ at 90 MHz. The VSWR has dropped to 5.2:1.

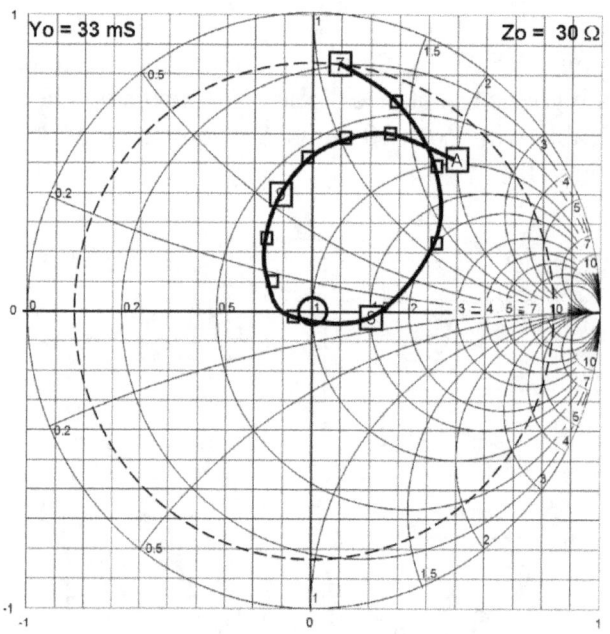

ROTATE PLOT - MOVE 62° ALONG LINE
FIGURE 10-10

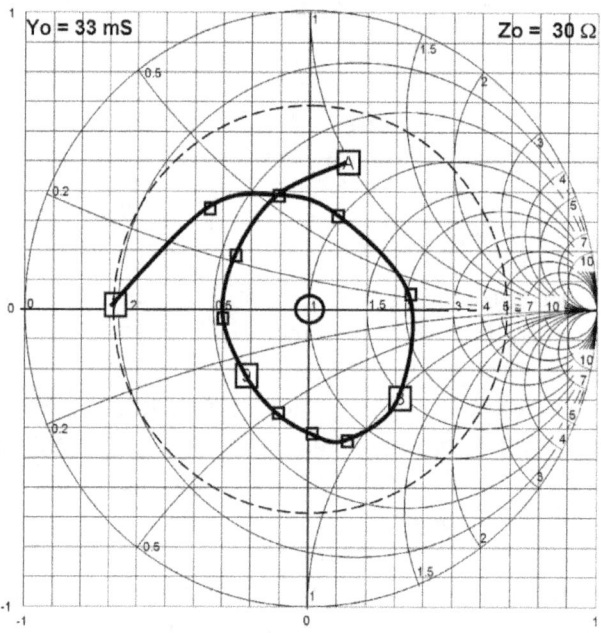

ADD SHUNT L - $b = -0.865$
FIGURE 10-11

Still there is need to move point *7* to a higher susceptance circle. Another rotation is in order. By moving further toward the radio gear on the coax 60° at 90 MHz produces Figure 10-12. The VSWR remains constant at 5.2:1.

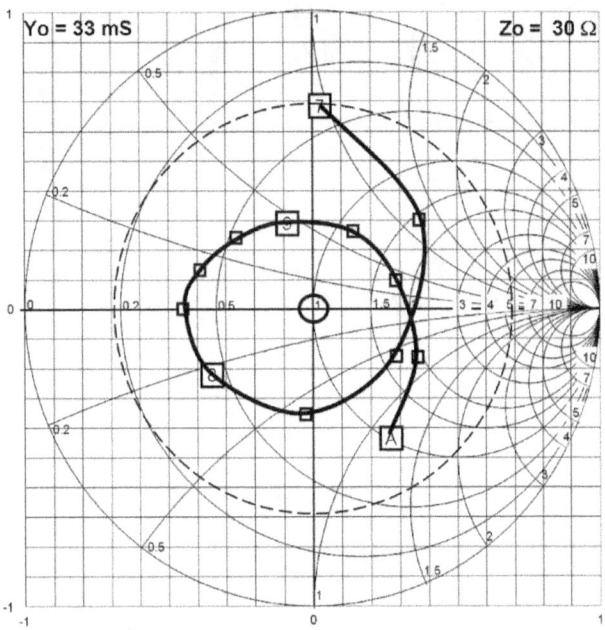

ROTATE PLOT - MOVE 60° ALONG LINE
FIGURE 10-12

Now a shunt stub should finish the procedure. However, point *7* will need an inductive compensation of $b = -1$ and this amount will pull some points below the real axis outward past the VSWR ring. A way to modify the situation is to add a shunt capacitor at this point, one with $b = +0.6$ at 90 MHz. Figure 10-13 (next page) shows the result: The VSWR has increased to 8.1:1 but a shunt stub will correct the situation.

With a final shunt stub the maximum VSWR has now been reduced to 3.2:1. This is an acceptable value for receiving antennas, but more is desired for transmitting antennas. The average transmitter available on today's market is of solid state design, and a 3:1 VSWR causes its internal circuitry to cut back its power output – a self-protection feature. However, many such transmitters incorporate an internal "antenna tuner" to compensate for VSWR up to 3:1 Note, however, that this device is *not* "tuning the antenna" as is a common misconception: it merely "tunes the *coax*" to deliver maximum power thereto.

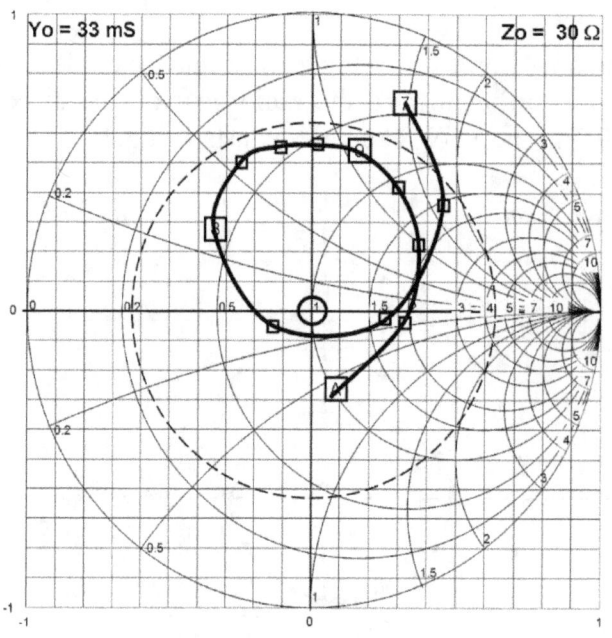

ADD SHUNT CAPACITOR
$b = +0.6$ @ 90 MHz
FIGURE 10-13

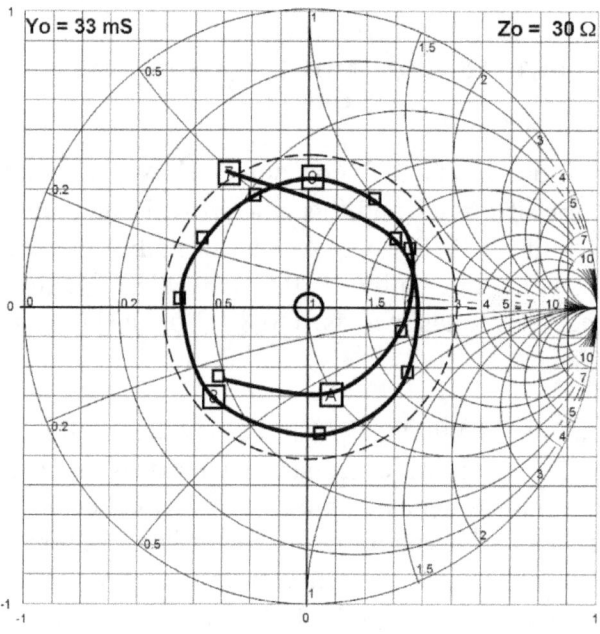

ADD SHUNT STUB
17 OHMS 80°
FIGURE 10-14

The completed matching system is shown in Figure 10-15.

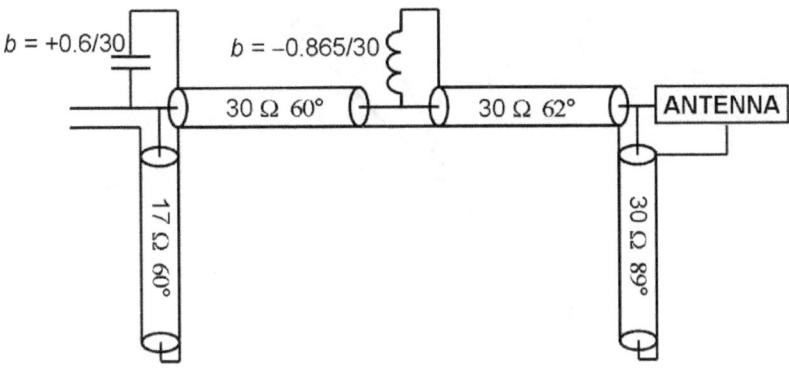

COMPLETE MATCHING SYSTEM
FIGURE 10-15

This matches a single monopole antenna to 30 ohms. A dipole will consist of two of these in series, to match to 60 ohms. As seen in Figure 10-14, a shift in Zo to 37.5 ohms would shift the entire pattern to the right a bit, and maintain it very closely inside the VSWR 3:1 circle. The dipole version would then become a 75 ohm balanced load. A simple half-octave cable balun as described in Chapter 7, Figure 7-4, would match it to ordinary 75 ohm cable.

Coax of Differing Zo

The analysis above speaks of 30 ohm and 17 ohm coax. This is easily assembled. Two pieces of coax tied in parallel, one of Zo = 75 ohms and the other of Zo = 50 ohms will synthesize coax of 30 ohms. And three pieces, each with Zo of 50 ohms, will synthesize 17.5 ohm coax. Thus the assemble of Figure 10-15 can be realized with standard, easily available coaxial cable.

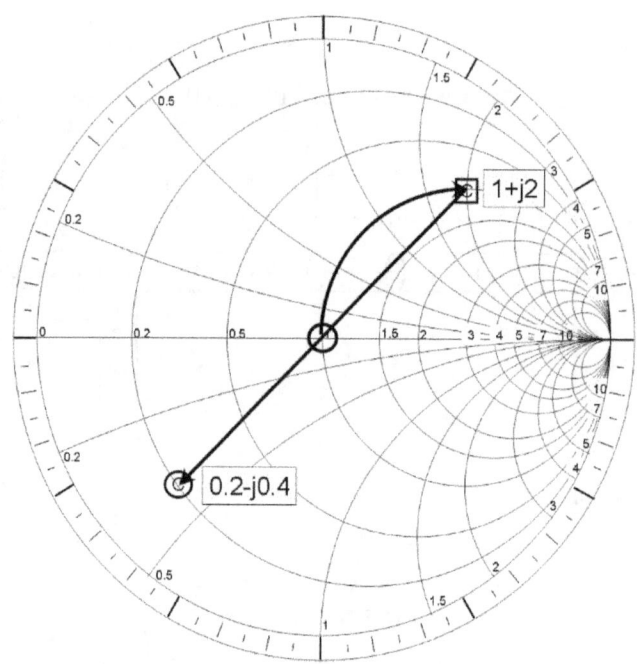

C=+j 2.0 ADMITTANCE AND IMPEDANCE
FIGURE 11-1

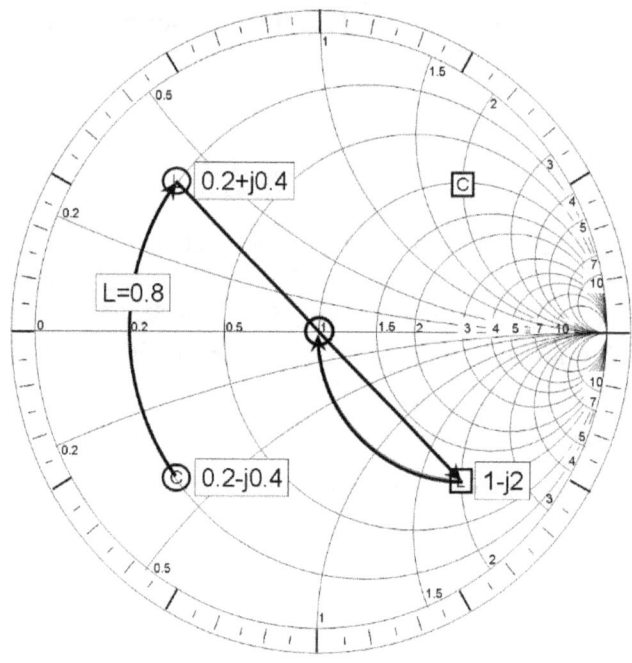

EFFECT OF SERIES L AND REMAINING C
FIGURE 11-2

Zo

A very short piece of coax open at the far end is a coaxial capacitor. Ordinary RG-8 or RG-58 cable has about 11 pF capacitance per foot. Likewise its conductor has about 27.5 nH per foot. These two determine the characteristic impedance of the line.

Oliver Heaviside analyzed such lines circa 1885. He described them as a chain of short segments, each with some inductance, resistance, capacitance, and leakage conductance. Here I ignore the resistance and conductance losses and show his depiction of a line as follows: this is a lossless line, a fair approximation. for radio work.

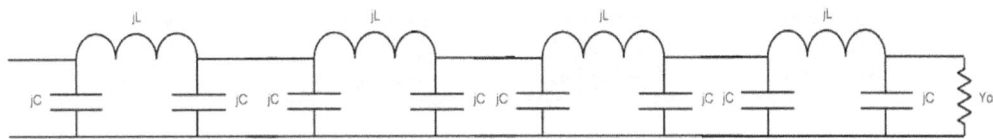

TRANSMISSION LINE MODEL (LOSSLESS)
FIGURE 11-3

With L and C properly set, conductance Yo appears at the left side of each C-L-C pi section, repeating everywhere along the line. Let's get a handle on this graphically with a smith chart. In Figure 11-1 the normalized conductance y_o sits at the center of the chart, by definition. Adding a shunt capacitor with normalized susceptance +j 2.0 pushes the admittance toward the upper right, to 1+j 2.0. Since we want to add a series reactance, switch to impedance mode by running half way 'round the chart, to the small circle in the lower left, where normalized impedance is found to be $z = 0.2 -j\ 0.4$

Refer now to Figure 11-2. If series inductance with normalized reactance +j 0.8 is added, and we convert back to admittance, we find that another identical capacitor brings the admittance at the left end of the first pi section back to y_o exactly. This action repeats all along the line. The normalized inductance per section is +j 0.8, and the total normalized capacity per section is +j 4.0, a ratio of ℓ/c of 0.8/4 = 0.2.

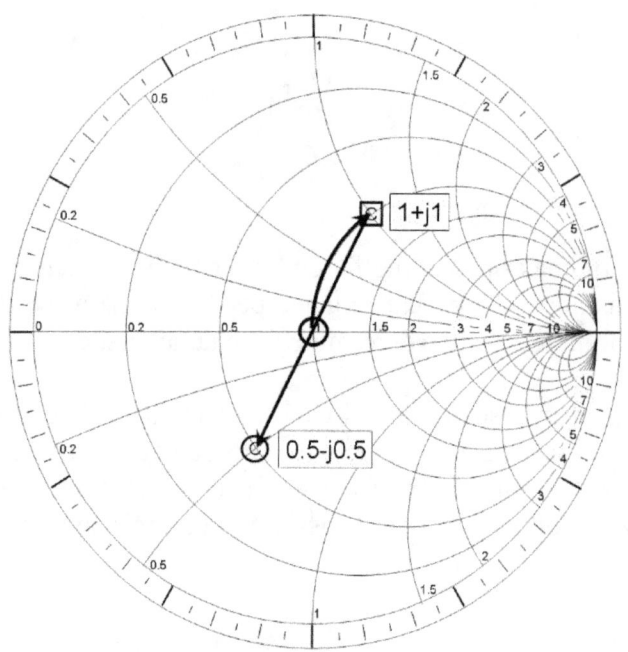

SITUATION AT HALF FREQUENCY – SHUNT C
FIGURE 11-4

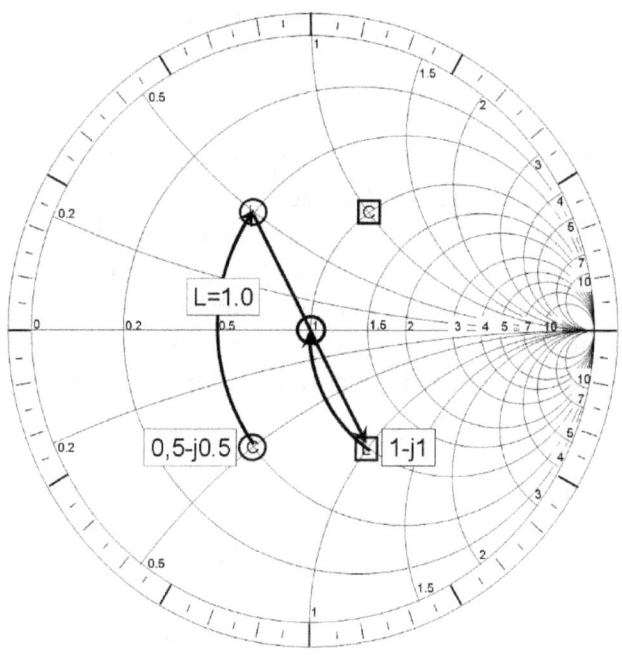

COMPLETING THE HALF FREQUENCY SITUATION
FIGURE 11-5

But if we drop to half frequency, where the susceptance of the capacitors and reactance of the inductor have half these values, things go awry. In Figure 11-4 the capacitance has value +j 1.0, and produces total impedance $z = 0.5 - j\, 0.5$. In Figure 11-5, an inductance with normalized value $-j\, 1.0$ followed by another shunt capacitance of susceptance $+j\, 1.0$ is now required to have y_o repeat at the left end of each pi section. The required ℓ/c ratio has risen to 0.5. It had been 0.2.

We continue this procedure, cutting the value of capacitance in half each time. The next iteration appears in Figure 11-6, with capacitance $b = +0.5$. The required inductance is now $+j\, 0.8$, and the normalized ℓ/c ratio has become 0.8, as in Figure 11-6.

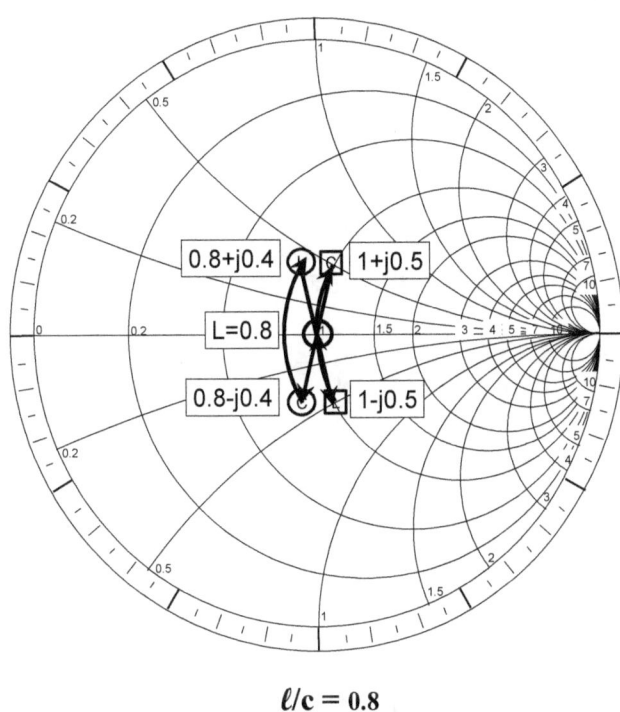

$\ell/c = 0.8$
FIGURE 11-6

We find that the ℓ/c ratio is rising, and in later figures it is seen to be approaching a limiting value of 1.0. In each case, c is a percentage of y_o: $c = k y_o$ and likewise ℓ is a percentage of z_o: $\ell = k z_o$, and as both become smaller and smaller, the value of k in both expressions becomes the same. Therefore we can say:

$$z_o = \ell/k\, y_o = c/k \qquad z_o = 1/y_o = k/c$$

and a result is that $z_o^2 = \ell/k \times k/c = \ell/c$. The characteristic impedance of the line is then just the square root of inductance/foot divided by capacitance/foot.

73

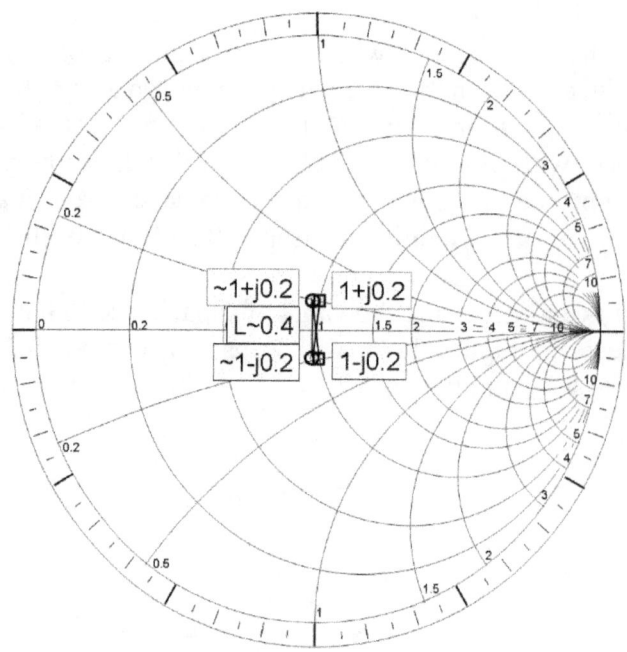

l/c ~ 1.0
FIGURE 11-7

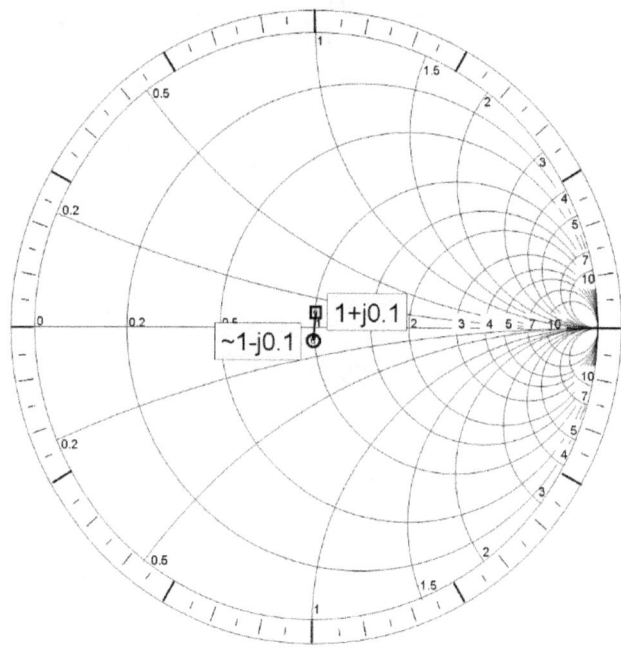

l/c ~ 1.0
FIGURE 11-8

NOTES

NOTES

APPENDIX A

PHASOR ARITHMETIC

Phasors are two-dimensional vectors, and complex numbers. They have two forms: magnitude and phase, such as $|Z| \angle \theta$; and real and Imaginary parts, such as $Z = R +j X$, as had been introduced in Chapter 1.

Magnitude

Magnitude, or size, of a phasor is the square root of the sum of squares of the real and imaginary parts.

$Z = R +j X$: $|Z| = \sqrt{(R^2 + X^2)}$
$Z^2 = (R^2 + X^2)$

Addition

Addition produces a new phasor whose real part is the sum of the two real parts, and ditto for the imaginary part.

$Z = Z_1 + Z_2$ $Z = R +j X = (R_1 + R_2) +j (X_1 + X_2)$

Inverse

The inverse of impedance Z is admittance Y. The real part is the real part of Z divided by the square of Z's magnitude, and ditto for the imaginary part, with a reversal of sign (positive becomes negative, and negative becomes positive.)

$Y = 1 / Z = G +j B = R / Z^2 \ -j X / Z^2$

Multiplication

The magnitude of the product is the product of the two magnitudes, and the phase angle is the sum of the two phase angles.

$Z = Z_1 Z_2 = |Z| \angle \theta = |Z_1| |Z_2| \angle (\theta_1 +\theta_2)$

Division

To divide, multiply by the inverse.

$$Z = Z_1 / Z_2 = Z_1 Y_2$$

These basic operations permit arithmetic manipulation of impedances and admittances.

Equality

A complex equation is actually two equations in one. The real parts are equal, and the imaginary parts are equal.

$$Z_1 = R_1 + X_1$$
$$Z_2 = R_2 + X_2$$

$Z_1 = Z_2$ so $R_1 = R_2$ and $X_1 = X_2$

NOTES

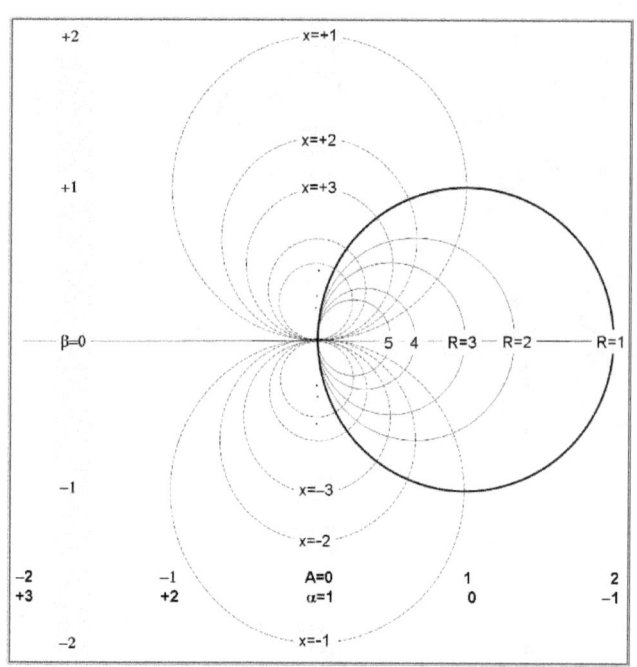

A – β PLOT
FIGURE B-1

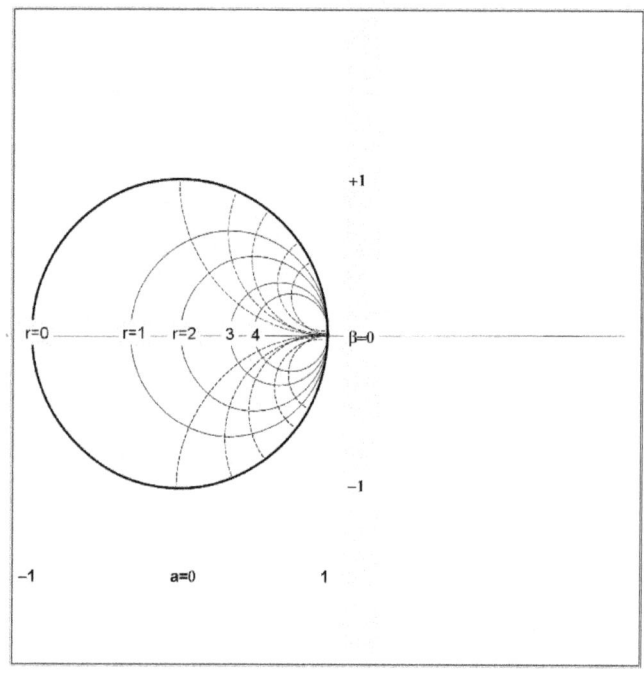

α – β PLOT
FIGURE B-2

APPENDIX B

THE $z - \rho$ CONFORMAL MAPPING
SMITH CHART MATH

The reflection coefficient is defined as

$$\rho = (z-1)/(z+1) \quad \text{so} \quad \rho = 1 - 2/(z+1)$$

Both ρ and z are complex numbers: $\quad \rho = \alpha + j\beta, \quad z = r + jx$

Then
$$\alpha + j\beta = 1 - 2/(r + jx + 1)$$
$$2/(\underline{r+1} + jx) = 1 - \alpha - j\beta$$
$$2/(R + jx) = A - j\beta$$
$$2 = (R + jx)(A - j\beta)$$

$$AR + \beta x = 2$$
$$Ax - \beta R = 0$$

$$A^2R + A\beta x = 2A$$
$$A\beta x - \beta^2 R = 0$$
$$(A^2 + \beta^2)R = 2A$$

$$A\beta R + \beta^2 x = 2\beta$$
$$A^2 x - A\beta R = 0$$
$$(A^2 + \beta^2)x = 2\beta$$

$$A^2 + \beta^2 = 2A/R$$
$$A^2 + \beta^2 = 2\beta/x$$

$$(A - 1/R)^2 + \beta^2 = 1/R^2$$
$$(\beta - 1/x)^2 + A^2 = 1/x^2$$

The first is the equation of a circle, radius $1/R$, centered at $A = 1/R$, $\beta = 0$.
The second defines another circle, radius $1/x$, centered at $\beta = 1/x$, $A = 0$.
These are plotted in Figure B-1.

Using the definitions of **A** and **R**, the first family is centered at $(1-\alpha) = 1/(r+1)$, or $\alpha = r/(r+1)$, $\beta = 0$, diameter $1/(r+1)$. The second family is centered at $\beta = 1/x$, $\alpha = 1$, diameter $1/x$. These circles are plotted in Figure B-2.

NOTES

www.ingramcontent.com/pod-product-compliance
Lightning Source LLC
Chambersburg PA
CBHW082244300426

44110CB00036B/2445